Vectors and Vector Operators

Student Monographs in Physics

Series Editor: Professor Douglas F Brewer
Professor of Experimental Physics, University of Sussex

Other books in the series:

Microcomputers
D G C Jones

Maxwell's Equations and their Applications
E G Thomas and A J Meadows

Oscillations and Waves
R Buckley

Fourier Transforms in Physics
D C Champeney

Kinetic Theory
J M Pendlebury

Diffraction
C Taylor

The Structures of Crystals
A M Glazer

Vectors and Vector Operators

P G Dawber

School of Mathematical and Physical Sciences, University of Sussex

Adam Hilger, Bristol

British Library Cataloguing in Publication Data

Dawber, P. G.
Vectors and vector operators——(Student monographs in physics)
1. Vector analysis 2. Mathematical physics
I. Title II. Series
515′.63′02453 QC20.7.V4

ISBN 0-85274-585-0

Published under the Adam Hilger imprint by IOP Publishing Limited
Techno House, Redcliffe Way, Bristol BS1 6NX, England

Typeset by KEYTEC, Bridport, Dorset
Printed in England by WBC Print Ltd, Barton Manor, St Philips, Bristol BS2 0RL

Contents

Preface

This book aims to give an introduction to the use of vectors and vector operators to first-year physics students many of whom have had little or no introduction to vectors at school and find their use a source of difficulty in their early physics courses. Most of the introductory texts on this subject are aimed at mathematicians, or written by mathematicians, and concentrate on the development of skills in the manipulation of vector relationships (e.g. showing that $(\boldsymbol{A} \wedge \boldsymbol{B}) \wedge (\boldsymbol{C} \wedge \boldsymbol{D}) = \ldots$). This book takes an essentially different approach in tying the development of vector techniques to problems in physics, showing that the vector form of physical equations is *simpler* both in terms of the physical space required, and also, much more importantly, in that the essential geometrical relationships between the quantities involved become immediately apparent. For this reason the use of the component form of vector equations has been avoided, except in a few instances where it simplifies the derivation of a particular result, and stress is placed on the fact that the vector form is independent of the choice of coordinate system. No attempt has been made to ensure that the proofs of the more difficult mathematical theorems are rigorous, since attention to this sort of detail often causes the beginner in this area to flounder.

Vector Algebra 1

1.1 Description of Vectors

Probably the first time we meet vectors is when we describe the positions of places. For example, we say that town B is 10 km north east of A. The vector joining B to A has length 10 km and a definite direction NE. Thus to describe positions in a two-dimensional plane we need two parameters, the length or magnitude and the direction. We could be more precise in specifying the direction by giving the angle between the direction of the vector to B and the direction North, so that NE becomes 45°, E becomes 90°, SE becomes 135° etc.

If we wish to specify the relative positon of an aeroplane we would need to give one more piece of information, the height. Thus, in three dimensions, we need three parameters to specify the vector position of the aeroplane. To simplify the discussion initially we will work with vectors in a two-dimensional plane but the extension to three dimensions will be very straightforward.

In two dimensions we can represent a vector by a straight line in the plane, the length of the line being the magnitude of the vector (drawn to some suitable scale) and its direction being specified by reference to a chosen direction for north.

The origin of the vector is not specified, so that all vectors with the same length and direction are the same vector e.g. the vectors $\boldsymbol{a}$ in figure 1.1.

Here we have introduced a notation $\boldsymbol{a}$; a bold-faced letter to represent the vector. An alternative way to describe a vector is to label the two end points e.g. A and B in figure 1.1 and write the vector

$$\boldsymbol{a} = \overrightarrow{\mathrm{AB}}.$$

Two different points C and D with the same relative positions will however specify the same vector so that

$$\boldsymbol{a} = \overrightarrow{\mathrm{AB}} = \overrightarrow{\mathrm{CD}}.$$

(In hand written notes it is customary to denote a vector by underlining e.g. $\boldsymbol{a}$ is written $\underline{a}$.)

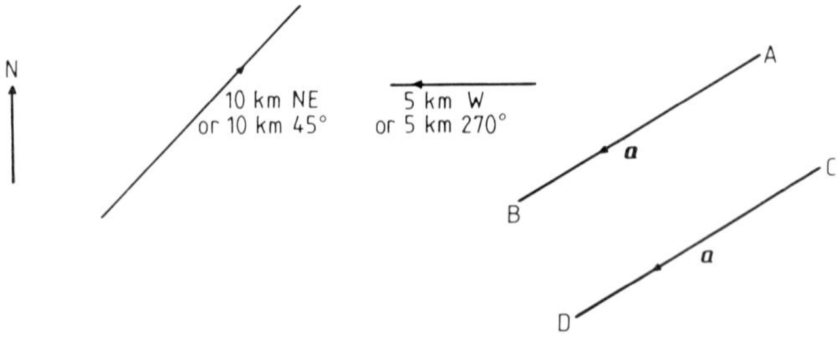

Figure 1.1

In physics we meet many variables which are vectors. Thus, apart from the position of a particle which is clearly a vector, as shown above, the velocity $\boldsymbol{v}$ is also a vector quantity, as it has both magnitude (e.g. 5 m s^{-1}) and direction (e.g. SE). Other vector quantities which you will probably have met already are the acceleration $\boldsymbol{a}$, the force $\boldsymbol{F}$ on a particle and the electric and magnetic fields $\boldsymbol{E}$ and $\boldsymbol{B}$.

Quantities which only have a magnitude and no direction are called scalars. Examples of scalars are the mass m and the kinetic energy $\frac{1}{2}mv^2$, of a particle. The magnitude of the vector $\boldsymbol{a}$ is simply the scalar length of $\boldsymbol{a}$ and is written as a (not bold face) or sometimes, to be more specific, as $|\boldsymbol{a}|$.

1.2 Scaling, Addition and Subtraction of Vectors

1.2.1 Scaling

We can multiply a vector by a scalar and thus change its length but not its direction e.g. $3\boldsymbol{a}$, $5\boldsymbol{a}$, $-\frac{1}{2}\boldsymbol{a}$, etc in figure 1.2.

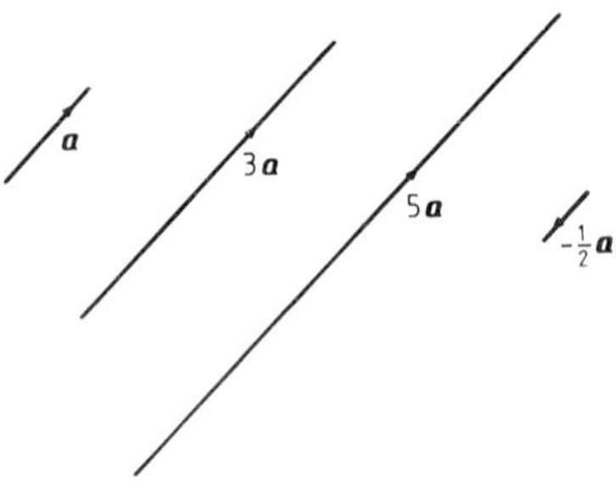

Figure 1.2

It is also convenient to introduce the concept of a unit vector. A vector of length 1 in the direction of the vector $\boldsymbol{a}$ is called a unit vector and is denoted $\hat{\boldsymbol{a}}$. It is the vector $\boldsymbol{a}$ scaled by the factor $1/a$, i.e. $\hat{\boldsymbol{a}} = \boldsymbol{a}/a$.

1.2.2 Addition

If we go from A to B by travelling 10 km NE and then from B to C by travelling 5 km W, we can find the position of C relative to A by drawing the two vectors $\overrightarrow{AB}$ and $\overrightarrow{BC}$ to scale on the paper as in figure 1.3(*a*). In this way we can construct the vector $\overrightarrow{AC}$ which is the VECTOR SUM of the two vectors $\overrightarrow{AB}$ and $\overrightarrow{BC}$. This can be written as the vector equation

$$\overrightarrow{AC} = \overrightarrow{AB} + \overrightarrow{BC} \qquad \text{or } \boldsymbol{c} = \boldsymbol{a} + \boldsymbol{b}.$$

We can continue this process to form the vector sum of several vectors, e.g. $\overrightarrow{AE} = \overrightarrow{AB} + \overrightarrow{BC} + \overrightarrow{CD} + \overrightarrow{DE}$ in figure 1.3(*b*).

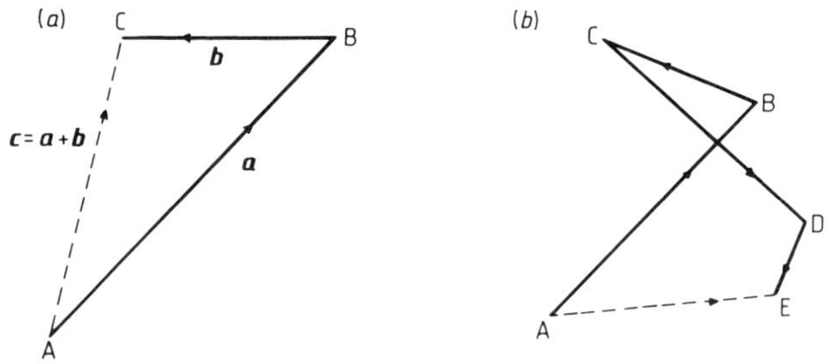

Figure 1.3

1.2.3 Subtraction

In order to subtract two vectors as in $\boldsymbol{c} = \boldsymbol{a} - \boldsymbol{b}$ we rewrite this equation as $\boldsymbol{c} = \boldsymbol{a} + (-\boldsymbol{b})$ and hence add the vector $-\boldsymbol{b}$ to $\boldsymbol{a}$ to produce $\boldsymbol{c}$ as in figure 1.4.

From the definition of addition it immediately follows that addition is commutative, i.e. $\boldsymbol{a} + \boldsymbol{b} = \boldsymbol{b} + \boldsymbol{a}$, as we can see in figure 1.5, and it is clear that $(c + d)\boldsymbol{a} = c\boldsymbol{a} + d\boldsymbol{a}$ and that $c(\boldsymbol{a} + \boldsymbol{b}) = c\boldsymbol{a} + c\boldsymbol{b}$.

In physics the rules for the addition of velocities, forces, etc are exactly those we have described above confirming our description of them as vectors.

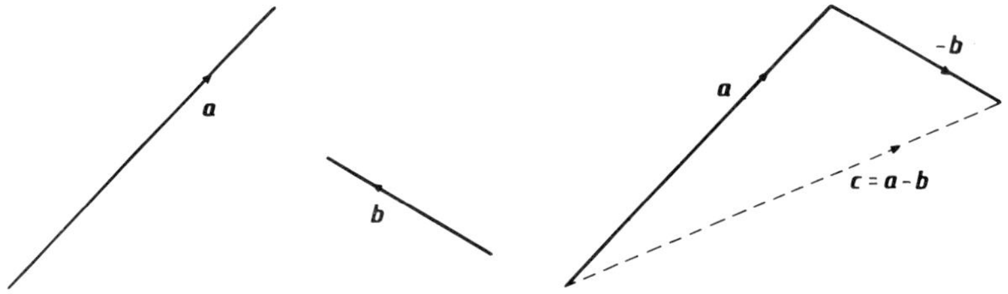

Figure 1.4

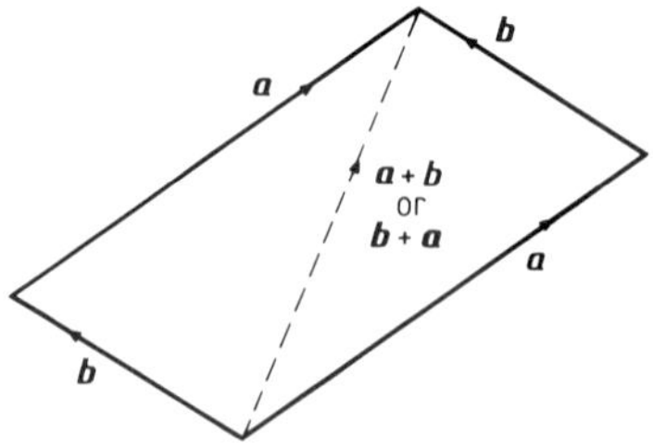

Figure 1.5

1.3 Decomposition and Resolution of Vectors

Since we can add together two vectors $\boldsymbol{a}$ and $\boldsymbol{b}$ to give a third vector $\boldsymbol{c} = \boldsymbol{a} + \boldsymbol{b}$, we can also do the reverse, i.e. we can write a vector $\boldsymbol{c}$ as the sum of two other vectors $\boldsymbol{a}$ and $\boldsymbol{b}$. Clearly the choice for $\boldsymbol{a}$ and $\boldsymbol{b}$ is infinite. In fact, if we choose two arbitrary non-parallel vectors $\boldsymbol{e}$ and $\boldsymbol{f}$ in a plane, then any vector in this plane can be written in the form

$$\boldsymbol{a} = c_1\boldsymbol{e} + c_2\boldsymbol{f}$$

where c_1 and c_2 are scalars, i.e. just numbers. This is illustrated in figure 1.6 where, by completing the parallelogram ADBC, we can see that

$$\boldsymbol{a} = \overrightarrow{\mathrm{AB}} = \overrightarrow{\mathrm{AD}} + \overrightarrow{\mathrm{DB}} = \overrightarrow{\mathrm{AD}} + \overrightarrow{\mathrm{AC}} = c_1\boldsymbol{e} + c_2\boldsymbol{f}$$

where $c_1 = \mathrm{AD}/\mathrm{AE}$ and $c_2 = \mathrm{AC}/\mathrm{AF}$.

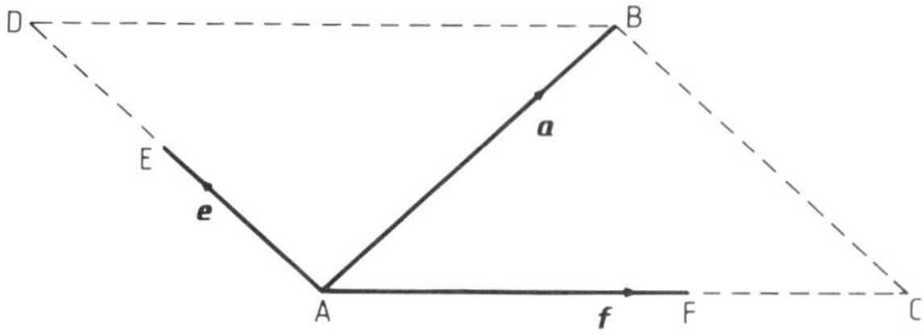

Figure 1.6

It is conventional to introduce Cartesian axes, $0X$ and $0Y$, to be two directions at right angles to one another in the plane, as in figure 1.7.

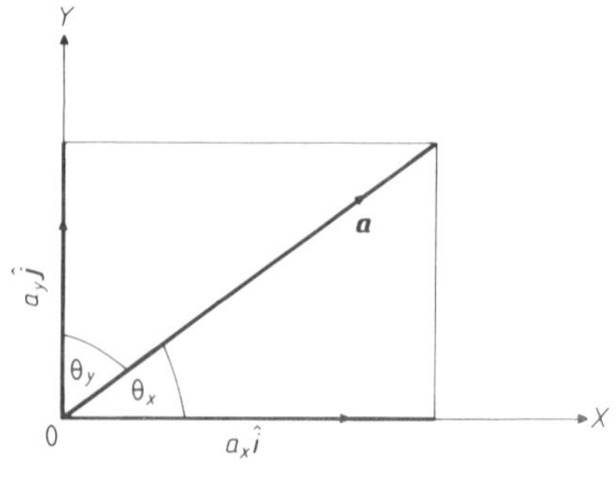

Figure 1.7

We can now select the two vectors $\boldsymbol{e}$ and $\boldsymbol{f}$ in the decomposition described above to be unit vectors $\hat{\boldsymbol{i}}$ and $\hat{\boldsymbol{j}}$ along the Cartesian axes $0X$ and $0Y$ so that

$$\boldsymbol{a} = a_x\hat{\boldsymbol{i}} + a_y\hat{\boldsymbol{j}}. \tag{1.1}$$

The vector $\boldsymbol{a}$ is said to be resolved into Cartesian components $a_x\hat{\boldsymbol{i}}$ and $a_y\hat{\boldsymbol{j}}$ along the x and y axes respectively. This is shown in figure 1.7. The components have magnitudes

$$a_x = a\cos\theta_x \qquad a_y = a\cos\theta_y$$

where θ_x and θ_y represent the angles between $\boldsymbol{a}$ and the x and y axes respectively. By the theorem of Pythagoras

$$a_x^2 + a_y^2 = a^2.$$

We can now generalise our work to three dimensions, introducing a third unit vector $\hat{\boldsymbol{k}}$ along the z axis (at right angles to the x and y axes as in figure 1.8) so that equation (1.1) becomes

$$\boldsymbol{a} = a_x\hat{\boldsymbol{i}} + a_y\hat{\boldsymbol{j}} + a_z\hat{\boldsymbol{k}}$$

with

$$a_x^2 + a_y^2 + a_z^2 = a^2. \tag{1.2}$$

The vector equation $\boldsymbol{a} = 0$ is thus seen to be a statement that all three components of $\boldsymbol{a}$ must vanish. The 0 in this equation is strictly a vector but it is unusual to use bold face type for this null vector. The vector equation $\boldsymbol{a} = \boldsymbol{b}$ implies that $\boldsymbol{a} - \boldsymbol{b} = 0$ and hence that $(a_x - b_x)\hat{\boldsymbol{i}} + (a_y - b_y)\hat{\boldsymbol{j}} + (a_z - b_z)\hat{\boldsymbol{k}} = 0$. It is thus equivalent to the three scalar equations

$$a_x = b_x \qquad a_y = b_y \qquad a_z = b_z. \tag{1.3}$$

We will occasionally use the Cartesian component form (1.3) in manipulating vectors; but this form, as it depends on the choice of axes, has neither the simplicity nor the immediate physical significance of the vector relationship $\boldsymbol{a} = \boldsymbol{b}$, which is written down without a choice of axes.

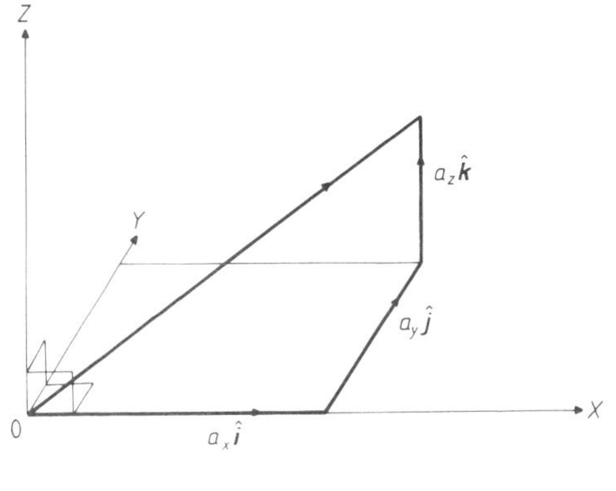

Figure 1.8

1.4 Examples of Vector Addition from Physics

Example 1.1

As mentioned in §1.1 forces are vector quantities. Thus, if a particle is subjected to several forces $\boldsymbol{F}_1, \boldsymbol{F}_2, \boldsymbol{F}_3, \ldots, \boldsymbol{F}_N$, the effect is the same as that of a single force $\boldsymbol{F}$ equal to the vector sum $\boldsymbol{F}_1 + \boldsymbol{F}_2 + \ldots + \boldsymbol{F}_N$. For example, we will calculate the acceleration (a vector) of a particle of mass 0.1 kg, free to move on a friction-free table, when subjected to a force of 3 N in the x direction and 4 N in the y direction. From figure 1.9, we see that the resultant force has a magnitude $|\boldsymbol{F}| = (F_1^2 + F_2^2)^{1/2} = 5$ N and acts in a direction φ relative to the x axis where $\tan\varphi = \frac{4}{3}$. By Newton's law, $\boldsymbol{F} = m\boldsymbol{a}$, the acceleration is in this same direction, with a magnitude $F/m = 50 \text{ m s}^{-2}$.

More interesting examples occur with the addition of velocities and we will discuss three more examples which, although all illustrating the same basic addition of vectors, are distinct physical problems.

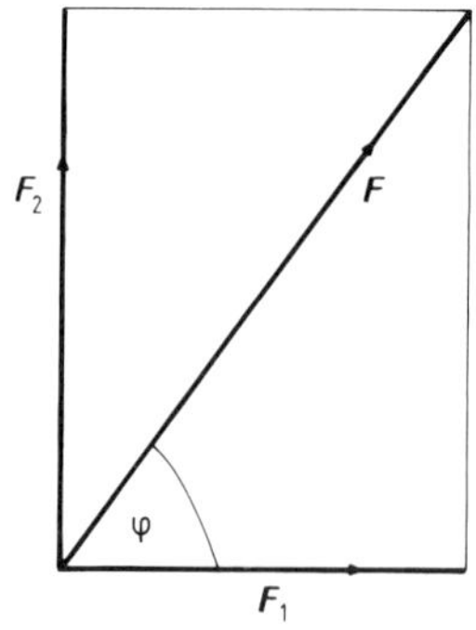

Figure 1.9

Example 1.2

First, we consider the effect of firing a gun from a moving object such as a tank. If the tank is travelling due north at 10 m s^{-1} and fires a shell at 200 m s^{-1} in a direction which appears to be due west to an observer on the tank, its true direction and speed relative to the ground will be given by the vector sum of these two velocities. From figure 1.10(a) we see that the true velocity has a magnitude $v = (200^2 + 10^2)^{1/2} = 205 \text{ m s}^{-1}$ and a direction which makes an angle φ north of due west where $\tan\varphi = 10/200$ i.e. $\varphi = \tan^{-1}0.05$.

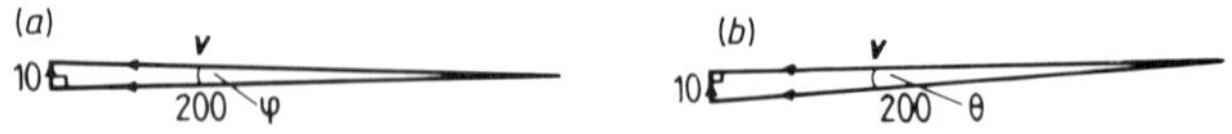

Figure 1.10

Example 1.3

A more relevant question is to ask in what direction should the gun be aimed to hit a target which is due west of the tank? In this case the gun must be fired in a direction θ south of due west so that when the velocity of the tank is added to that of the shell the resulting total velocity is in the due west direction. From figure 1.10(*b*) we see that the angle θ is given by $\sin\theta = 10/200$ so that $\theta = \sin^{-1}0.05$.

These examples show that, if an observer moving with velocity $\boldsymbol{v}_0$ sees an object moving with an apparent velocity $\boldsymbol{v}_a$ when its true velocity is $\boldsymbol{v}_t$, then

$$\boldsymbol{v}_t = \boldsymbol{v}_a + \boldsymbol{v}_0. \tag{1.4}$$

Example 1.4

A river flows with a speed 1 m s^{-1}. A boy wishes to swim across the river to a point directly opposite him on the far bank. If he can swim with a water speed of 2 m s^{-1}, at what angle θ should he aim relative to the bank? Here again we can use equation (1.4) by noting that the water speed is the speed of the boy relative to an observer moving with the water. Hence in the notation of equation (1.4) this is $\boldsymbol{v}_a$. The true velocity $\boldsymbol{v}_t$ must be directly across the river so the appropriate vector diagram is shown in figure 1.11.

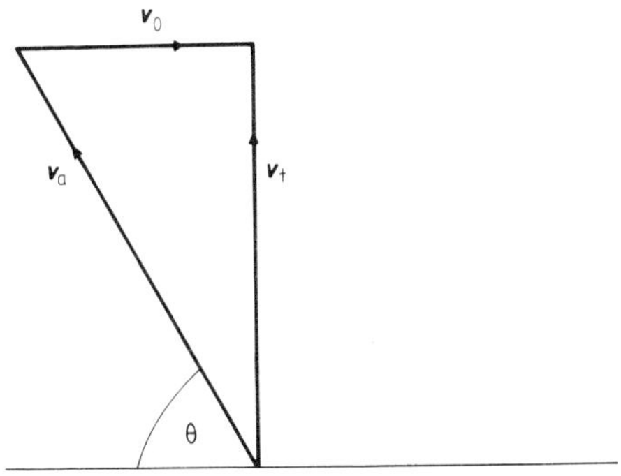

Figure 1.11

The angle θ is thus given by

$$\cos\theta = \tfrac{1}{2} \qquad \theta = 60°.$$

Products of Vectors 2

Having considered the sum of two vectors in the previous chapter we now wish to move on to the multiplication of vectors. In defining the sum, in §1.2, we were guided by our 'experimental' knowledge of the result of adding two vectors representing the positions of places in a two-dimensional plane. We found that this definition also described correctly the law of addition of other physically significant vector quantities such as velocity and force. When we come to the definition of a product the same is not true. We can in fact choose different ways of defining a product. In practice two definitions produce results of physical significance. The scalar product, written $\boldsymbol{a}\cdot\boldsymbol{b}$, is a way of forming a scalar quantity from two vectors $\boldsymbol{a}$ and $\boldsymbol{b}$, and the vector product, written $\boldsymbol{a}\wedge\boldsymbol{b}$, is a way of forming a vector quantity. We consider these two types of product separately below.

2.1 The Scalar Product

The scalar product $\boldsymbol{a}\cdot\boldsymbol{b}$ of two vectors $\boldsymbol{a}$ and $\boldsymbol{b}$ is defined to be the scalar quantity

$$\boldsymbol{a}\cdot\boldsymbol{b} = |\boldsymbol{a}|\,|\boldsymbol{b}|\cos\theta \tag{2.1}$$

where θ is the angle between the two vectors $\boldsymbol{a}$ and $\boldsymbol{b}$ when they are drawn from the same origin as in figure 2.1.

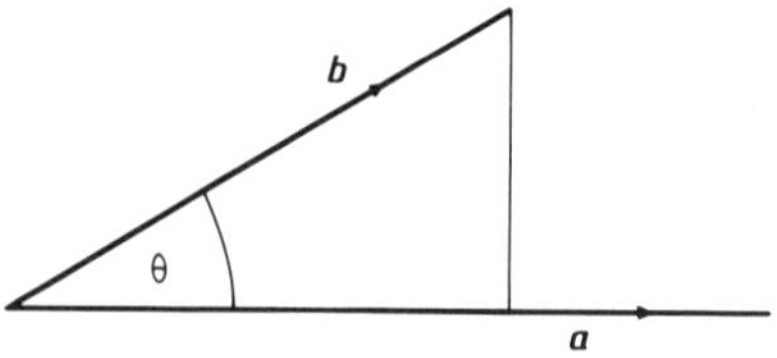

Figure 2.1

We can also write the result as ax (projection of $\boldsymbol{b}$ on $\boldsymbol{a}$) or bx (projection of $\boldsymbol{a}$ on $\boldsymbol{b}$), demonstrating the obvious symmetry of the result

$$\boldsymbol{a}\cdot\boldsymbol{b} = \boldsymbol{b}\cdot\boldsymbol{a}. \tag{2.2}$$

The scalar product is associative i.e.

$$\boldsymbol{a}\cdot(\boldsymbol{b} + \boldsymbol{c}) = \boldsymbol{a}\cdot\boldsymbol{b} + \boldsymbol{a}\cdot\boldsymbol{c}$$

as can be seen in figure 2.2 where the projection of $(\boldsymbol{b} + \boldsymbol{c})$ on $\boldsymbol{a}$ is equal to the sum of the projections of $\boldsymbol{b}$ and $\boldsymbol{c}$ on $\boldsymbol{a}$.

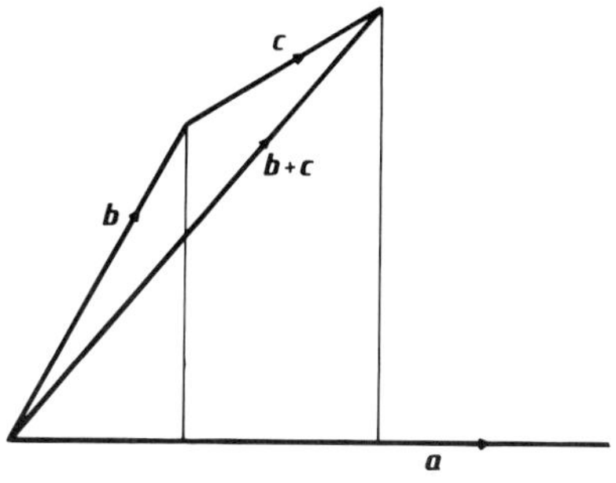

Figure 2.2

We note that the scalar product $\boldsymbol{a}\cdot\boldsymbol{b}$ is zero if $\boldsymbol{a}$ and $\boldsymbol{b}$ are perpendicular so that $\boldsymbol{a}\cdot\boldsymbol{b} = 0$ does not necessarily imply that either $\boldsymbol{a}$ or $\boldsymbol{b}$ is zero.

It is useful to note that the scalar products of the Cartesian unit vectors are

$$\hat{\boldsymbol{i}}\cdot\hat{\boldsymbol{i}} = \hat{\boldsymbol{j}}\cdot\hat{\boldsymbol{j}} = \hat{\boldsymbol{k}}\cdot\hat{\boldsymbol{k}} = 1$$

and

$$\hat{\boldsymbol{i}}\cdot\hat{\boldsymbol{j}} = \hat{\boldsymbol{j}}\cdot\hat{\boldsymbol{k}} = \hat{\boldsymbol{k}}\cdot\hat{\boldsymbol{i}} = 0. \tag{2.3}$$

These relations enable us to write the scalar product of two vectors $\boldsymbol{a}$ and $\boldsymbol{b}$ in terms of their Cartesian components as

$$\begin{aligned}\boldsymbol{a}\cdot\boldsymbol{b} &= (a_x\hat{\boldsymbol{i}} + a_y\hat{\boldsymbol{j}} + a_z\hat{\boldsymbol{k}})\cdot(b_x\hat{\boldsymbol{i}} + b_y\hat{\boldsymbol{j}} + b_z\hat{\boldsymbol{k}}) \\ &= a_xb_x + a_yb_y + a_zb_z. \end{aligned} \tag{2.4}$$

The scalar product described above appears naturally in many physical problems and we look at some examples here.

2.1.1 *Physical Applications of the Scalar Product*

The work done, W, by a force $\boldsymbol{F}$ acting on a body, when its point of application moves a distance $\boldsymbol{d}$, is the product of the displacement and the component of the force in the direction of the displacement (or alternatively, the product of the force and the component of the displacement in the direction of the force). From figure 2.3 this is

$$W = (F\cos\theta)d = Fd\cos\theta = \boldsymbol{F}\cdot\boldsymbol{d}. \tag{2.5}$$

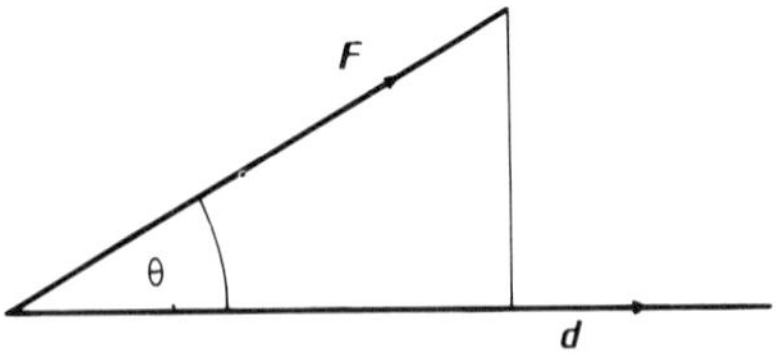

Figure 2.3

Example 2.1
As an example we will calculate the work done per second by a horse which pulls a barge along a canal as it walks along the towpath, as in figure 2.4.

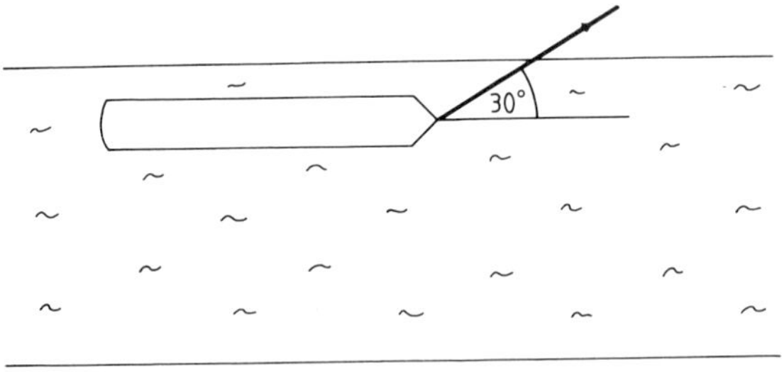

Figure 2.4

If the horse walks at 1 $m\,s^{-1}$, the tension in the tow rope is 300 N, and the angle between the rope and the direction of the canal is 30°, how much work is done by the horse per second (i.e. what is the power produced by the horse)? The work done, W, is given by

$$W = \boldsymbol{F}\cdot\boldsymbol{d} = Fd\cos\theta$$

where d is the distance moved by the barge in one second

i.e. $$W = 300 \times 1 \times \cos 30° = 150\ \mathrm{J\,s^{-1}}\ (\text{or W}).$$

When work is done on a body it gives rise to an increase in the energy of the body. This energy may be kinetic (if the body accelerates) or potential (i.e. energy stored due to the change of position) or it may be dissipated in doing work against frictional forces. Thus energy in any form is a scalar quantity and in many cases the potential energy of a system is written as a scalar product of vector quantities. Two important examples are the potential energy of an electric dipole $\boldsymbol{p}$ in an electric field $\boldsymbol{E}$

$$V = -\boldsymbol{p}\cdot\boldsymbol{E} \tag{2.6}$$

and the potential energy of a magnetic dipole of moment $\boldsymbol{\mu}$ in a magnetic field $\boldsymbol{B}$

$$V = -\boldsymbol{\mu}\cdot\boldsymbol{B}. \tag{2.7}$$

We illustrate the second of these results in the following example.

Example 2.2
A square loop of wire of side L carrying a current i is placed in a magnetic field $\boldsymbol{B}$ with the plane of the loop parallel to the field. If the loop is pivoted about an axis perpendicular to the field, as in figure 2.5, it experiences a torque (as in a galvanometer). Calculate the work done by the field if the loop is allowed to turn until its plane is perpendicular to the field.

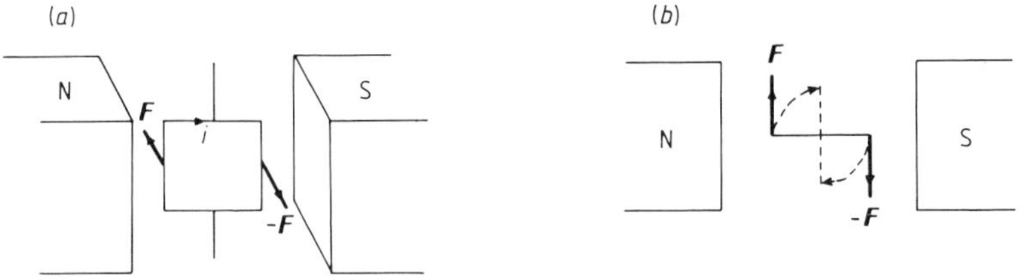

Figure 2.5

We solve this problem in two different ways, first using equation (2.5), then using equation (2.7).

The force on a wire of length L carrying a current i in a field B is BiL in a direction given by Fleming's left-hand rule. Thus the forces acting on two of the sides of the loop give rise to the torque as in figure 2.5(b). As the loop rotates these each move a distance $L/2$ in the direction of the force so that the work done is

$$W = 2FL/2 = BiL^2. \qquad (2.8)$$

Alternatively, the stored potential energy V is given by equation (2.7), and the change in V will be related to the work W by

$$W = -(\text{change in } V) = V(\text{initial}) - V(\text{final}).$$

The magnetic moment $\boldsymbol{\mu}$ has a magnitude given by $\mu = iA$ where A is the area of the loop and has a direction perpendicular to the loop as shown in figure 2.6. It starts perpendicular to $\boldsymbol{B}$ and finishes parallel to $\boldsymbol{B}$. Thus we have

$$W = 0 - (-iAB) = iL^2B$$

which agrees with the result (2.8).

Figure 2.6

2.2 The Vector Product

The vector product of two vectors $\boldsymbol{a}$ and $\boldsymbol{b}$ is written in the form $\boldsymbol{a}\wedge\boldsymbol{b}$ (sometimes $\boldsymbol{a}\times\boldsymbol{b}$) and is defined to be the vector with magnitude $|\boldsymbol{a}|\,|\boldsymbol{b}|\sin\theta$ where θ is the angle between the two vectors $\boldsymbol{a}$ and $\boldsymbol{b}$ drawn from the same origin ($\theta < 180°$). The direction of $\boldsymbol{a}\wedge\boldsymbol{b}$ is along an axis perpendicular to the plane containing $\boldsymbol{a}$ and $\boldsymbol{b}$ with the sense chosen to be that of a right-handed screw rotated through θ from $\boldsymbol{a}$ to $\boldsymbol{b}$ about that axis. This is illustrated in figure 2.7.

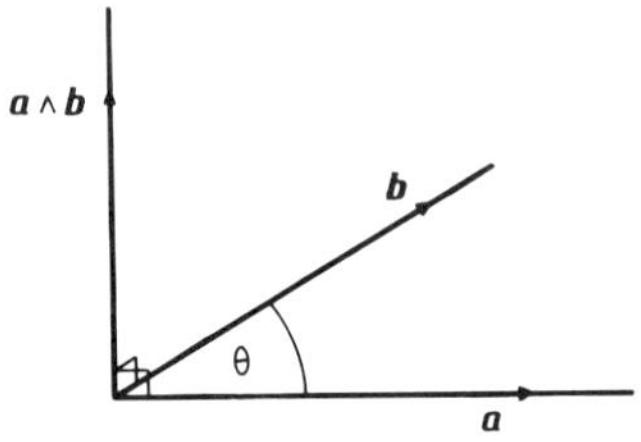

Figure 2.7

From its definition we see that this vector product is not commutative. In fact

$$\boldsymbol{a}\wedge\boldsymbol{b} = -\boldsymbol{b}\wedge\boldsymbol{a} \tag{2.9}$$

since the vector $\boldsymbol{b}\wedge\boldsymbol{a}$ has the same magnitude as $\boldsymbol{a}\wedge\boldsymbol{b}$ but the opposite direction, as we must now rotate from $\boldsymbol{b}$ to $\boldsymbol{a}$ via θ.

The magnitude of $\boldsymbol{a}\wedge\boldsymbol{b}$, i.e. $|\boldsymbol{a}\wedge\boldsymbol{b}|$, is equal to the area of the parallelogram formed with $\boldsymbol{a}$ and $\boldsymbol{b}$ as two adjacent sides.

The Cartesian unit vectors described in §1.3 satisfy the relations

$$\hat{\boldsymbol{i}}\wedge\hat{\boldsymbol{i}} = \hat{\boldsymbol{j}}\wedge\hat{\boldsymbol{j}} = \hat{\boldsymbol{k}}\wedge\hat{\boldsymbol{k}} = 0 \tag{2.10}$$

since the vector product of any vector with itself vanishes as $\theta = 0$, but we also have

$$\begin{aligned} \hat{\boldsymbol{i}}\wedge\hat{\boldsymbol{j}} &= -\hat{\boldsymbol{j}}\wedge\hat{\boldsymbol{i}} = \hat{\boldsymbol{k}} \\ \hat{\boldsymbol{j}}\wedge\hat{\boldsymbol{k}} &= -\hat{\boldsymbol{k}}\wedge\hat{\boldsymbol{j}} = \hat{\boldsymbol{i}} \\ \hat{\boldsymbol{k}}\wedge\hat{\boldsymbol{i}} &= -\hat{\boldsymbol{i}}\wedge\hat{\boldsymbol{k}} = \hat{\boldsymbol{j}}. \end{aligned} \tag{2.11}$$

Note that we can obtain the second and third relations above from the first by cyclically permuting the vectors $\hat{\boldsymbol{i}}$, $\hat{\boldsymbol{j}}$ and $\hat{\boldsymbol{k}}$, i.e. by simultaneously changing $\hat{\boldsymbol{i}}\to\hat{\boldsymbol{j}}$, $\hat{\boldsymbol{j}}\to\hat{\boldsymbol{k}}$, $\hat{\boldsymbol{k}}\to\hat{\boldsymbol{i}}$. This useful property will hold for any vector relationship writen in terms of the Cartesian base vectors.

From equation (2.11) we can write the vector product of two vectors in terms of their Cartesian components

$$\boldsymbol{a}\wedge\boldsymbol{b} = (a_x\hat{\boldsymbol{i}} + a_y\hat{\boldsymbol{j}} + a_z\hat{\boldsymbol{k}})\wedge(b_x\hat{\boldsymbol{i}} + b_y\hat{\boldsymbol{j}} + b_z\hat{\boldsymbol{k}})$$
$$= (a_yb_z - b_ya_z)\hat{\boldsymbol{i}} + (a_zb_x - b_za_x)\hat{\boldsymbol{j}} + (a_xb_y - b_xa_y)\hat{\boldsymbol{k}}. \quad (2.12)$$

We do not wish to emphasise this result in component form since the expression on the right contains no obvious information about the size and direction of the vector $\boldsymbol{a}\wedge\boldsymbol{b}$. However it is occasionally useful in the derivation of specific results. Before leaving equation (2.12) we will also point out that it may be written in the form

$$\boldsymbol{a}\wedge\boldsymbol{b} = \begin{vmatrix} \hat{\boldsymbol{i}} & \hat{\boldsymbol{j}} & \hat{\boldsymbol{k}} \\ a_x & a_y & a_z \\ b_x & b_y & b_z \end{vmatrix} \quad (2.13)$$

where the right-hand side is a determinant.

We now illustrate the immense importance of the vector product in physical applications with some examples.

2.2.1 *Physical Applications of the Vector Product*

(a) Moment of a force

In elementary mechanics, the moment of a force (or torque) acting on a rigid body, which can rotate about an axis perpendicular to a plane containing the force, is defined to be the magnitude of the force multiplied by the perpendicular distance from the force to the axis.

This is shown in figure 2.8(a) from which we see that the moment $\boldsymbol{M}$ has a magnitude $|\boldsymbol{F}|p$, or $Fr\sin\theta$. The moment also has a direction which is along the axis in a direction such that the action of the force is that of a right-handed screw (out of the paper in figure 2.8). These results can be

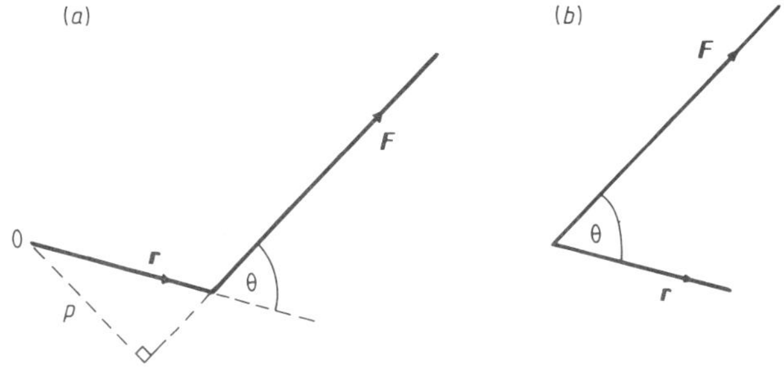

Figure 2.8

summarised in the vector equation

$$\boldsymbol{M} = \boldsymbol{r} \wedge \boldsymbol{F}. \tag{2.14}$$

(Note that, to get the direction, we should draw $\boldsymbol{r}$ and $\boldsymbol{F}$ from a common origin as in figure 2.8(*b*) and chose θ to be the angle less than 180°, as in the definition at the beginning of this section.)

The definition (2.14) of the moment of a force, or torque, at the origin, is in fact quite general. However, if the axis which passes through the origin is not in a direction perpendicular to the plane containing $\boldsymbol{r}$ and $\boldsymbol{F}$, then $\boldsymbol{M}$ will not be parallel to the axis and we must take the component of $\boldsymbol{M}$ in the direction of the axis as the moment (or torque) about that axis.

(b) Angular momentum

A particle of mass m moving with velocity $\boldsymbol{v}$ in a plane has an angular momentum $\boldsymbol{L}$, about an axis passing through the origin, given by the expression

$$\boldsymbol{L} = m\boldsymbol{r} \wedge \boldsymbol{v} \tag{2.15}$$

where $\boldsymbol{r}$ is the position vector of the particle relative to the origin. Since the linear momentum of the particle is defined to be $\boldsymbol{p} = m\boldsymbol{v}$, we can write this as

$$\boldsymbol{L} = \boldsymbol{r} \wedge \boldsymbol{p}. \tag{2.16}$$

(c) Force on a charged particle in a magnetic field

The force on a particle of charge q, moving in a magnetic field $\boldsymbol{B}$ (the Lorentz force) is given by

$$\boldsymbol{F} = q\boldsymbol{v} \wedge \boldsymbol{B}. \tag{2.17}$$

(d) Torque on a dipole in a field

The torque on an electric dipole $\boldsymbol{p}$ in an electric field $\boldsymbol{E}$ is given by

$$\boldsymbol{T} = \boldsymbol{p} \wedge \boldsymbol{E}. \tag{2.18}$$

This result can easily be seen by taking the dipole to be two charges $+q$ and $-q$ separated by a small distance $2d$ as in figure 2.9.

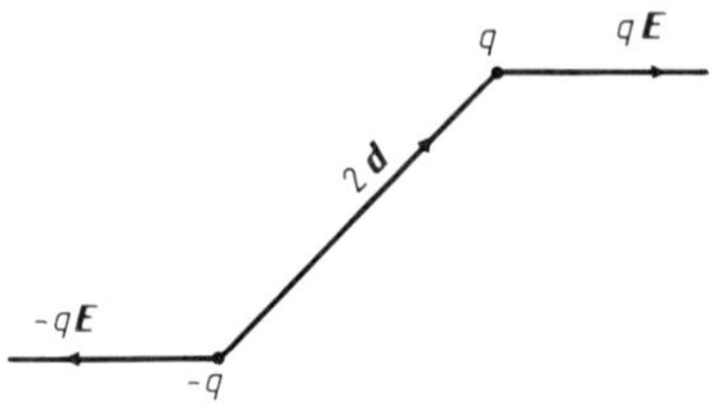

Figure 2.9

The force on each charge has a magnitude qE. The resulting moment is given by equation (2.14) as

$$\boldsymbol{M} = \boldsymbol{T} = 2\boldsymbol{d}\wedge(q\boldsymbol{E}) = 2q\boldsymbol{d}\wedge\boldsymbol{E}.$$

The electric dipole moment is $\boldsymbol{p} = 2\boldsymbol{d}q$ so this is the same as (2.18).

A similar result holds for the torque on a magnetic dipole in a magnetic field $\boldsymbol{B}$:

$$\boldsymbol{T} = \boldsymbol{\mu}\wedge\boldsymbol{B}. \tag{2.19}$$

2.3 Products of Three or More Vectors

Having defined products of two vectors in the previous two sections we can go on to construct products of three or more vectors. In particular, two different triple products yielding either a scalar or a vector occur frequently and we now give their definitions.

(a) The triple scalar product $\boldsymbol{a}\cdot\boldsymbol{b}\wedge\boldsymbol{c}$

This is the scalar product between the vector $\boldsymbol{b}\wedge\boldsymbol{c}$ and the vector $\boldsymbol{a}$. Since the scalar product of two vectors is independent of the order it follows that

$$\boldsymbol{a}\cdot\boldsymbol{b}\wedge\boldsymbol{c} = \boldsymbol{b}\wedge\boldsymbol{c}\cdot\boldsymbol{a}. \tag{2.20}$$

This has the immediate interpretation as the volume of the parallelepiped with edges $\boldsymbol{a}$, $\boldsymbol{b}$ and $\boldsymbol{c}$, drawn from the same origin, as can be seen from figure 2.10. The area of the base is $A = |\boldsymbol{a}\wedge\boldsymbol{b}|$. The volume is $V = Ah$, where h is

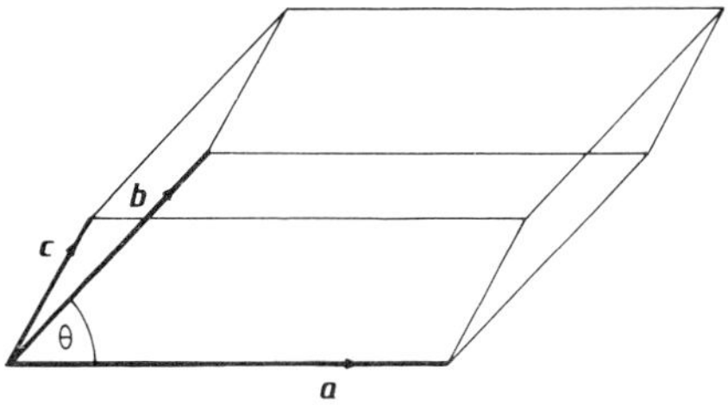

Figure 2.10

the height and can be written as $h = \boldsymbol{c}\cdot\hat{\boldsymbol{n}}$ where $\hat{\boldsymbol{n}}$ is a unit vector perpendicular to the base. Clearly $\boldsymbol{a}\wedge\boldsymbol{b} = A\hat{\boldsymbol{n}}$ and thus $V = A\boldsymbol{c}\cdot\hat{\boldsymbol{n}} = |\boldsymbol{a}\wedge\boldsymbol{b}\cdot\boldsymbol{c}|$. Since this volume does not depend on which face we call the base it follows that

$$\boldsymbol{a}\wedge\boldsymbol{b}\cdot\boldsymbol{c} = \boldsymbol{b}\wedge\boldsymbol{c}\cdot\boldsymbol{a} = \boldsymbol{c}\wedge\boldsymbol{a}\cdot\boldsymbol{b}. \tag{2.21}$$

Combining this result with (2.20) we see that the triple scalar product depends only on the cyclic order *abc* and not on the position of the $\cdot$ and $\wedge$ in the product i.e.

$$\boldsymbol{a}\wedge\boldsymbol{b}\cdot\boldsymbol{c} = \boldsymbol{b}\wedge\boldsymbol{c}\cdot\boldsymbol{a} = \boldsymbol{a}\cdot\boldsymbol{b}\wedge\boldsymbol{c}. \tag{2.22}$$

The sign is reversed if the cyclic order is broken by permuting two of the vectors.

Although we usually avoid using component form, we may express the triple vector product as

$$\boldsymbol{a}\cdot\boldsymbol{b}\wedge\boldsymbol{c} = a_x(b_yc_z - b_zc_y) + a_y(b_zc_x - b_xc_z) + a_z(b_xc_y - b_yc_x) \tag{2.23}$$

or in terms of the determinantal expression (2.13)

$$\boldsymbol{a}\cdot\boldsymbol{b}\wedge\mathbf{c} = \begin{vmatrix} a_x & a_y & a_z \\ b_x & b_y & b_z \\ c_x & c_y & c_z \end{vmatrix}. \tag{2.24}$$

(b) The triple vector product $\boldsymbol{a}\wedge(\boldsymbol{b}\wedge\boldsymbol{c})$

This is the vector product between the vector $\boldsymbol{a}$ and the vector $\boldsymbol{b}\wedge\boldsymbol{c}$. The result is a vector in the plane containing $\boldsymbol{b}$ and $\boldsymbol{c}$ (since $\boldsymbol{b}\wedge\boldsymbol{c}$ is perpendicular to this plane and $\boldsymbol{a}\wedge(\boldsymbol{b}\wedge\boldsymbol{c})$ is perpendicular to $\boldsymbol{b}\wedge\boldsymbol{c}$). In fact by geometrical considerations or by writing out the Cartesian components in detail it is easily shown that

$$\boldsymbol{a}\wedge(\boldsymbol{b}\wedge\boldsymbol{c}) = (\boldsymbol{a}\cdot\boldsymbol{c})\boldsymbol{b} - (\boldsymbol{a}\cdot\boldsymbol{b})\boldsymbol{c}. \tag{2.25}$$

The position of the brackets in $\boldsymbol{a}\wedge(\boldsymbol{b}\wedge\boldsymbol{c})$ is clearly of vital importance.

Differentiation of Vectors

3

3.1 Definition of the Derivative

So far we have considered constant vectors, but in order to discuss the dynamics of particle motion we will need to consider time-dependent vectors and their rate of change, i.e. we need to be able to differentiate vectors. The derivative of a vector is defined in exactly the same way as for a scalar. For example, suppose we have a vector $\boldsymbol{r}$ which is a function of time i.e. $\boldsymbol{r}(t)$. The time derivative is

$$\frac{\mathrm{d}\boldsymbol{r}}{\mathrm{d}t} = \lim_{\delta t \to 0} \frac{\boldsymbol{r}(t + \delta t) - \boldsymbol{r}(t)}{\delta t}. \tag{3.1}$$

Notice that this derivative, being the difference between two vectors, is itself a vector which in general will not be parallel to $\boldsymbol{r}$. A limiting case is motion in a circle about the origin, where the direction of $\boldsymbol{r}$ changes continuously but its magnitude remains constant. In this case $\mathrm{d}\boldsymbol{r}/\mathrm{d}t$ is a vector perpendicular to $\boldsymbol{r}(t)$. In terms of the components along stationary Cartesian axes, we note that

$$\frac{\mathrm{d}\boldsymbol{r}}{\mathrm{d}t} = \frac{\mathrm{d}x}{\mathrm{d}t}\hat{\boldsymbol{i}} + \frac{\mathrm{d}y}{\mathrm{d}t}\hat{\boldsymbol{j}} + \frac{\mathrm{d}z}{\mathrm{d}t}\hat{\boldsymbol{k}}. \tag{3.2}$$

From the basic definition (3.1) it follows that

$$\begin{aligned}
\frac{\mathrm{d}(\boldsymbol{a} + \boldsymbol{b})}{\mathrm{d}t} &= \frac{\mathrm{d}\boldsymbol{a}}{\mathrm{d}t} + \frac{\mathrm{d}\boldsymbol{b}}{\mathrm{d}t} \\
\frac{\mathrm{d}[c(t)\boldsymbol{a}]}{\mathrm{d}t} &= \frac{\mathrm{d}c(t)}{\mathrm{d}t}\boldsymbol{a} + c(t)\frac{\mathrm{d}\boldsymbol{a}}{\mathrm{d}t} \\
\frac{\mathrm{d}(\boldsymbol{a}\cdot\boldsymbol{b})}{\mathrm{d}t} &= \boldsymbol{a}\cdot\frac{\mathrm{d}\boldsymbol{b}}{\mathrm{d}t} + \frac{\mathrm{d}\boldsymbol{a}}{\mathrm{d}t}\cdot\boldsymbol{b} \\
\frac{\mathrm{d}(\boldsymbol{a}\wedge\boldsymbol{b})}{\mathrm{d}t} &= \boldsymbol{a}\wedge\frac{\mathrm{d}\boldsymbol{b}}{\mathrm{d}t} + \frac{\mathrm{d}\boldsymbol{a}}{\mathrm{d}t}\wedge\boldsymbol{b}.
\end{aligned} \tag{3.3}$$

In the above work we have taken t to represent the time, but everything we have said applies when the vectors are functions of any scalar variable t.

We can also extend the usual definition of integration to cover the case of vector functions of scalar variables. For instance, if we write $\boldsymbol{v}(t) = \mathrm{d}\boldsymbol{r}(t)/\mathrm{d}t$, then the inverse is

$$\int_{t_1}^{t_2} \boldsymbol{v}(t)\,\mathrm{d}t = \int_{t_1}^{t_2} \mathrm{d}\boldsymbol{r}(t) = \boldsymbol{r}(t_2) - \boldsymbol{r}(t_1). \tag{3.4}$$

This vector equation could be written in component form as three equations each of which is of the usual form for one-dimensional motion.

3.2 Particle Dynamics

If the position of a particle relative to the origin at time t is described by the vector $\boldsymbol{r}(t)$, we can obtain the velocity $\boldsymbol{v}(t)$ and the acceleration $\boldsymbol{a}(t)$ by differentiating with respect to time. The velocity is

$$\boldsymbol{v}(t) = \frac{\mathrm{d}\boldsymbol{r}(t)}{\mathrm{d}t} \tag{3.5}$$

and the acceleration is

$$\boldsymbol{a}(t) = \frac{\mathrm{d}\boldsymbol{v}(t)}{\mathrm{d}t} = \frac{\mathrm{d}^2\boldsymbol{r}(t)}{\mathrm{d}t^2}. \tag{3.6}$$

These equations can be integrated to give the velocity and position of a particle at time t in terms of the acceleration and time. Integrating equation (3.6) gives

$$\int_0^T \mathrm{d}\boldsymbol{v} = \int_0^T \boldsymbol{a}(t)\,\mathrm{d}t$$

or

$$\boldsymbol{v}(T) - \boldsymbol{v}(0) = \int_0^T \boldsymbol{a}(t)\,\mathrm{d}t.$$

It is customary to call the initial velocity $\boldsymbol{u}$ so this equation becomes

$$\boldsymbol{v}(T) = \boldsymbol{u} + \int_0^T \boldsymbol{a}(t)\,\mathrm{d}t. \tag{3.7}$$

For the special, but important, case of a constant acceleration $\boldsymbol{a}(t) = \boldsymbol{a}$ this equation becomes

$$\boldsymbol{v}(T) = \boldsymbol{u} + \boldsymbol{a}T \tag{3.8}$$

which is the vector equivalent of the usual scalar equation $v = u + at$. Similarly we can integrate equation (3.5) to get

$$\int_0^T \mathrm{d}\boldsymbol{r}(t) = \int_0^T \boldsymbol{v}(t)\,\mathrm{d}t$$

or

$$\boldsymbol{r}(T) - \boldsymbol{r}(0) = \int_0^T \boldsymbol{v}(t)\,\mathrm{d}t. \tag{3.9}$$

If we make the usual assumption that we start at $\boldsymbol{r} = 0$ when $t = 0$, this becomes

$$\boldsymbol{r}(T) = \int_0^T \boldsymbol{v}(t)\,\mathrm{d}t.$$

For the special case of a constant acceleration, we can use the result (3.8) to give

$$\begin{aligned}\boldsymbol{r}(T) &= \int_0^T (\boldsymbol{u} + \boldsymbol{a}t)\,\mathrm{d}t \\ &= \boldsymbol{u}T + \tfrac{1}{2}\boldsymbol{a}T^2 \qquad (3.10)\end{aligned}$$

which is the vector equivalent of the usual scalar equation $s = ut + \frac{1}{2}at^2$.

3.3 Plane Polar Coordinates and Circular Motion

There are many problems in physics where the symmetry makes the use of polar coordinates preferable to the Cartesian system introduced in §1.3. In two dimensions the position of a point relative to the origin is specified by the magnitude of $\boldsymbol{r}$ and the angle θ relative to the x-axis, as shown in figure 3.1. (This is essentially the same as the description of places on a map but the choice of the angle is different in that $\theta = 0$ is the x-axis and θ is positive for an anticlockwise rotation from this axis.)

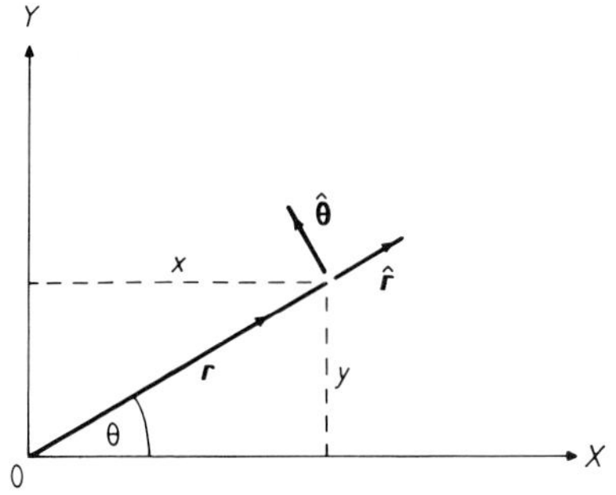

Figure 3.1

The relationship between the two sets of coordinates is easily seen from figure 3.1 to be

$$\begin{aligned} x &= r\cos\theta \qquad & r &= (x^2 + y^2)^{1/2} \\ y &= r\sin\theta \qquad & \theta &= \tan^{-1}(y/x). \end{aligned} \qquad (3.11)$$

In the Cartesian system we introduced the unit vectors $\hat{\boldsymbol{\imath}}$ and $\hat{\boldsymbol{\jmath}}$ and for the polar coordinates we define two unit vectors $\hat{\boldsymbol{r}}$ and $\hat{\boldsymbol{\theta}}$. $\hat{\boldsymbol{r}}$ is just a unit vector along the instantaneous direction of $\boldsymbol{r}$ (i.e. $\hat{\boldsymbol{r}} = \boldsymbol{r}/r$) and $\hat{\boldsymbol{\theta}}$ is a unit vector perpendicular to $\boldsymbol{r}$, in the sense shown in figure 3.1. There is an essential difference between these polar unit vectors and the Cartesian ones in that they are not fixed in direction. In fact, they are both functions of θ and rotate as θ varies. Strictly we should write $\hat{\boldsymbol{r}}(\theta)$ and $\hat{\boldsymbol{\theta}}(\theta)$ but this θ-dependence is not usually made explicit. If the position vector $\boldsymbol{r}$ is varying with some parameter such as time, then $\hat{\boldsymbol{r}}$ and $\hat{\boldsymbol{\theta}}$ will also vary with time (but always remain perpendicular to one another). In this case it is important to know the derivatives of $\hat{\boldsymbol{r}}$ and $\hat{\boldsymbol{\theta}}$. Consider a small time interval dt; θ will change by $d\theta$ and r by dr. From figure 3.2 we can see that

$$d\boldsymbol{r} = dr\boldsymbol{r} + rd\theta\hat{\boldsymbol{\theta}}. \tag{3.12}$$

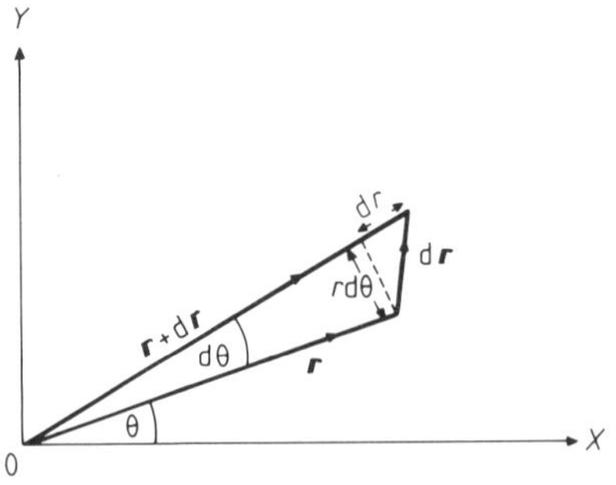

Figure 3.2

If we now take the special case where $\boldsymbol{r}$ is the unit vector $\hat{\boldsymbol{r}}$ with length 1, then $dr = 0$ and $d\hat{\boldsymbol{r}} = d\theta\hat{\boldsymbol{\theta}}$, i.e. the change in $\hat{\boldsymbol{r}}$ must be perpendicular to $\hat{\boldsymbol{r}}$. We thus have

$$\dot{\hat{\boldsymbol{r}}} = \frac{d\hat{\boldsymbol{r}}}{dt} = \frac{d\theta}{dt}\hat{\boldsymbol{\theta}} = \dot{\theta}\hat{\boldsymbol{\theta}}. \tag{3.13}$$

In a similar way we see that $\hat{\boldsymbol{\theta}}$ will change in direction at the same rate to remain perpendicular to $\hat{\boldsymbol{r}}$, so that

$$\dot{\hat{\boldsymbol{\theta}}} = \frac{d\hat{\boldsymbol{\theta}}}{dt} = -\dot{\theta}\hat{\boldsymbol{r}}. \tag{3.14}$$

3.3.1 *Uniform Circular Motion*

The special case of uniform circular motion, where a particle is rotating about the origin with constant angular velocity $\dot{\theta} = \omega$ and constant radial distance r, is of particular importance, so we note that in this case we have

$$\boldsymbol{r} = r\hat{\boldsymbol{r}}.$$

The velocity is given by

$$\begin{aligned} \boldsymbol{v} = \dot{\boldsymbol{r}} &= \dot{r}\hat{\boldsymbol{r}} + r\dot{\hat{\boldsymbol{r}}} \\ &= 0 + r\dot{\theta}\hat{\boldsymbol{\theta}} \text{ from equation (3.13)} \\ &= r\omega\hat{\boldsymbol{\theta}}. \end{aligned} \tag{3.15}$$

The acceleration is

$$\begin{aligned} \boldsymbol{a} = \dot{\boldsymbol{v}} = r\omega\dot{\hat{\boldsymbol{\theta}}} &= r\omega(-\dot{\theta}\hat{\boldsymbol{r}}) \text{ from equation (3.14)} \\ &= -r\omega^2\hat{\boldsymbol{r}}. \end{aligned} \tag{3.16}$$

This is the usual centripetal acceleration $r\omega^2$ directed towards the origin. Notice how simply this centripetal acceleration is derived using these vector techniques and also that the direction of the acceleration is specified.

3.3.2 *General Case*

If we remove the constraints applied above and allow ω and r to vary with time, the more general form for the velocity and acceleration can be obtained by differentiating the equation $\boldsymbol{r} = r\hat{\boldsymbol{r}}$ as above giving

$$\begin{aligned} \boldsymbol{v} &= \dot{r}\hat{\boldsymbol{r}} + r\dot{\hat{\boldsymbol{r}}} \\ &= \dot{r}\hat{\boldsymbol{r}} + r\omega\hat{\boldsymbol{\theta}} \end{aligned} \tag{3.17}$$

and the acceleration

$$\begin{aligned} \boldsymbol{a} = \dot{\boldsymbol{v}} &= \ddot{r}\hat{\boldsymbol{r}} + \dot{r}\omega\hat{\boldsymbol{\theta}} + \dot{r}\omega\hat{\boldsymbol{\theta}} + r\dot{\omega}\hat{\boldsymbol{\theta}} - r\omega^2\hat{\boldsymbol{r}} \\ &= (\ddot{r} - r\omega^2)\hat{\boldsymbol{r}} + (2\dot{r}\omega + r\dot{\omega})\hat{\boldsymbol{\theta}}. \end{aligned} \tag{3.18}$$

In deriving these equations we have used the general results given in equations (3.13) and (3.14). The terms in equation (3.18) are: $\ddot{r}\hat{\boldsymbol{r}}$ an acceleration in the radial direction arising from the change in the radial speed, $-r\omega^2\hat{\boldsymbol{r}}$ the centripetal acceleration in the radial direction which we met in equation (3.16) for uniform rotation, $r\dot{\omega}\hat{\boldsymbol{\theta}}$ an acceleration in the tangential direction $\hat{\boldsymbol{\theta}}$ due to the change in the angular velocity, and finally $2\dot{r}\omega\hat{\boldsymbol{\theta}}$, half of which arises from the change in the direction of the radial velocity and half of which comes from the change in tangential velocity due to the change in the radial distance.

3.3.3 *The Vector Angular Velocity*

It is frequently convenient to introduce the vector angular velocity $\boldsymbol{\omega}$ which is a vector of magnitude ω in the direction perpendicular to the plane of motion in the same sense as the angular momentum vector (see equation (2.15)). In terms of $\hat{\boldsymbol{r}}$ and $\hat{\boldsymbol{\theta}}$ we thus have

$$\boldsymbol{\omega} = \omega\hat{\boldsymbol{r}} \wedge \hat{\boldsymbol{\theta}}.$$

The vector $\boldsymbol{\omega}\wedge\boldsymbol{r}$ is then in the direction $\hat{\boldsymbol{\theta}}$. Using (2.25)

$$\begin{aligned}\boldsymbol{\omega}\wedge\boldsymbol{r} &= \omega r(\hat{\boldsymbol{r}}\wedge\hat{\boldsymbol{\theta}})\wedge\hat{\boldsymbol{r}}\\ &= \omega r\hat{\boldsymbol{\theta}}.\end{aligned}$$

From (3.15) this is just the velocity for a particle in uniform circular motion, and equation (3.17) for the more general case can be written in the alternative form

$$\boldsymbol{v} = \dot{r}\hat{\boldsymbol{r}} + \boldsymbol{\omega}\wedge\boldsymbol{r}. \tag{3.19}$$

The angular momentum (2.15) is given by

$$\begin{aligned}\boldsymbol{L} &= m\boldsymbol{r}\wedge\boldsymbol{v} = m\boldsymbol{r}\wedge(\dot{r}\hat{\boldsymbol{r}} + \boldsymbol{\omega}\wedge\boldsymbol{r})\\ &= mr^2\boldsymbol{\omega} \text{ (using (2.25)).}\end{aligned} \tag{3.20}$$

The scalar quantity mr^2 is called the 'moment of inertia' of the particle about this axis through the origin and is usually given the symbol I.

3.3.4 The Equation of Motion in Angular Form

The rate of change of the angular momentum is given by

$$\begin{aligned}\frac{\mathrm{d}\boldsymbol{L}}{\mathrm{d}t} &= m\frac{\mathrm{d}\boldsymbol{r}}{\mathrm{d}t}\wedge\boldsymbol{v} + m\boldsymbol{r}\wedge\frac{\mathrm{d}\boldsymbol{v}}{\mathrm{d}t}\\ &= m\boldsymbol{v}\wedge\boldsymbol{v} + \boldsymbol{r}\wedge\frac{\mathrm{d}\boldsymbol{p}}{\mathrm{d}t}\\ &= \boldsymbol{r}\wedge\boldsymbol{F}\\ &= \boldsymbol{M}\end{aligned} \tag{3.21}$$

where $\boldsymbol{F}$ is the force on the particle and $\boldsymbol{M}$ is the moment of this force (or torque) about the axis through the origin. This equation is the equivalent of Newton's law with the force replaced by the moment of the force $\boldsymbol{M}$ and the momentum replaced by the moment of momentum or angular momentum $\boldsymbol{L}$.

3.3.5 Rotational Motion of Rigid Bodies

Although we have derived equation (3.21) for a single particle it will also hold for a finite solid body provided that the angular momentum is defined as a sum over all the atoms in the solid, $\boldsymbol{L} = \Sigma_i m_i\boldsymbol{r}_i\wedge\boldsymbol{v}_i$, and the moment $\boldsymbol{M}$ as the sum over all external moments about the same origin, $\boldsymbol{M} = \Sigma_i\boldsymbol{r}_i\wedge\boldsymbol{F}_i$. (The contributions to $\boldsymbol{M}$ of all the internal forces cancel out as is shown in detail in mechanics texts.)

If 0 is some point fixed in a rigid body, then the instantaneous motion of the body is described by giving the velocity of 0 and the motion of the body relative to the point 0. This latter motion is, instantaneously, a rotation with angular velocity vector $\boldsymbol{\omega}$ passing through the point 0. Ignoring for the moment the motion of 0 the velocity of the atom at the point $\boldsymbol{r}_i$ relative to 0 will have magnitude $\omega r_i \sin\theta$ (see figure 3.3) in the direction perpendicular to both $\boldsymbol{\omega}$ and $\boldsymbol{r}$. We can thus write this velocity as

$$\dot{\boldsymbol{r}}_i = \boldsymbol{v}_i = \boldsymbol{\omega}\wedge\boldsymbol{r}_i. \tag{3.22}$$

(The velocity of 0 can be added to this result if 0 is moving.) The motion of the vector $\boldsymbol{r}_i$ about the axis $\boldsymbol{\omega}$ is called 'precession' and we often meet this type of motion in physics.

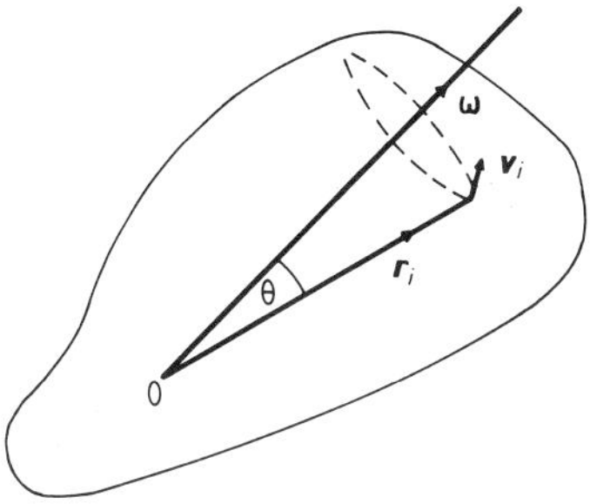

Figure 3.3

The generalisation of the result (3.20) for the angular momentum of the rigid body is thus

$$\begin{aligned}\boldsymbol{L} &= \sum_i m_i \boldsymbol{r}_i\wedge\boldsymbol{v}_i \\ &= \sum_i m_i \boldsymbol{r}_i\wedge(\boldsymbol{\omega}\wedge\boldsymbol{r}_i)\end{aligned}$$

and using (2.25) this becomes

$$\boldsymbol{L} = \sum_i m_i r_i^2\boldsymbol{\omega} - \sum_i m_i(\boldsymbol{r}_i\cdot\boldsymbol{\omega})\boldsymbol{r}_i.$$

In general, the second term on the right does not vanish and the angular momentum will not be parallel to the angular velocity vector! It is this result which makes the general motion of a rigid body a difficult subject. It can be shown that, for all bodies, a set of three principal axes can be found in the body so that rotation about these principal axes does yield the simple result $\boldsymbol{L} = I\boldsymbol{\omega}$ with $\boldsymbol{L}$ parallel to $\boldsymbol{\omega}$ and I one of the three principal moments of inertia. For bodies of high symmetry these axes will be along symmetry axes and we will confine our attention to such bodies.

This is not the place to go into detail about rigid-body motion but we hope we have shown in the derivation of the above equations that this subject is one where vector methods will be a powerful tool and we close this section by looking at one particular form of motion, precession, which is of great significance in physics. It is met in classical physics in connection with the motion of a spinning body and in quantum mechanics and atomic physics when discussing the effect of magnetic fields on atomic electrons. Although the electronic problem is somewhat easier to handle, we will look first at classical gyroscopic precession since this is the more familiar topic.

Consider a rigid body spinning about its axis of symmetry with angular velocity vector $\boldsymbol{\omega}$, such that one point 0 on the axis is fixed as in figure 3.4. Although we will find later that the axis precesses about the vertical direction we first assume that the angular momentum of the body is just that due to its spinning motion and write

$$\boldsymbol{L} = I\boldsymbol{\omega} \tag{3.23}$$

where I is the principal moment of inertia about the axis of symmetry. The external moment acting on the body is due to the gravitational force and has a value $mgl \sin\theta$, where l is the distance from the centre of gravity of the body to the origin and θ the angle between the axis $\boldsymbol{\omega}$ and the vertical z-axis (see figure 3.4). The direction of this moment is that of the vector $\hat{z} \wedge \boldsymbol{\omega}$ so that we can write it vectorially as

$$\boldsymbol{M} = mgl\hat{z} \wedge \hat{\boldsymbol{\omega}}$$

where $\hat{\boldsymbol{\omega}}$ is a unit vector in the direction of $\boldsymbol{\omega}$. The equation of motion (3.21) is $\boldsymbol{M} = \mathrm{d}\boldsymbol{L}/\mathrm{d}t$ so that

$$mgl\hat{z} \wedge \hat{\boldsymbol{\omega}} = I\mathrm{d}\boldsymbol{\omega}/\mathrm{d}t. \tag{3.24}$$

From this equation we see that the change $\mathrm{d}\boldsymbol{\omega}$ is perpendicular to $\boldsymbol{\omega}$ so that the magnitude of $\boldsymbol{\omega}$ remains constant and only its direction changes. We can thus write equation (3.24) as

$$mgl\hat{z} \wedge \hat{\boldsymbol{\omega}} = I\omega\frac{\mathrm{d}\hat{\boldsymbol{\omega}}}{\mathrm{d}t}$$

or

$$\frac{\mathrm{d}\hat{\boldsymbol{\omega}}}{\mathrm{d}t} = \frac{mgl}{I\omega}(\hat{z} \wedge \hat{\boldsymbol{\omega}}). \tag{3.25}$$

This equation has exactly the same form as equation (3.22) with the unit

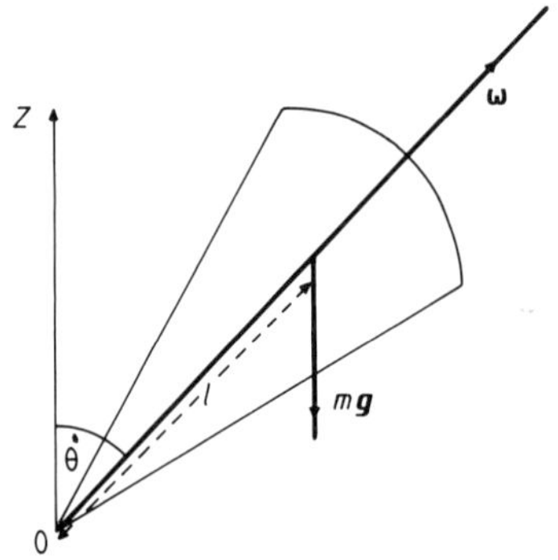

Figure 3.4

vector $\hat{\boldsymbol{\omega}}$ replacing $\boldsymbol{r}_i$ and $(mgl/I\omega)\hat{\boldsymbol{z}}$ replacing $\boldsymbol{\omega}$. It thus represents precession of the vector $\hat{\boldsymbol{\omega}}$ about the z-axis with precessional angular velocity $\Omega_{\mathrm{p}} = mgl/I\omega$.

Although we have obtained this solution using the approximation that we can ignore the precessional contribution to the total angular momentum, the form of the solution still holds if this is included. The precessional frequency is then found to be changed by an amount which is small, provided that Ω_{p} is much smaller than ω.

We now turn to the quantum mechanical problem. In classical physics a particle with charge q and angular momentum $\boldsymbol{L}$ has a magnetic moment given by $\boldsymbol{\mu} = q\boldsymbol{L}/2m$. In quantum mechanics the same result holds for the orbital motion of an electron in an atom but it is found that there is an additional magnetic moment associated with the 'intrinsic spin' of the electron. This intrinsic spin $\boldsymbol{S}$ is an internal angular momentum and the magnetic moment associated with it has twice the classical value, i.e. $\boldsymbol{\mu}_{\mathrm{s}} = -e\boldsymbol{S}/m$. In the presence of a magnetic field $\boldsymbol{B}$, the electron's magnetic moment experiences a torque or moment (see equation (2.19)) $\boldsymbol{M} = \boldsymbol{\mu}\wedge\boldsymbol{B}$, which must be equal to the rate of change of the angular momentum. Hence, ignoring any orbital effects, we have

$$\boldsymbol{M} = -\frac{e\boldsymbol{S}\wedge\boldsymbol{B}}{m} = \frac{\mathrm{d}\boldsymbol{S}}{\mathrm{d}t}. \qquad (3.26)$$

Without any approximation this equation has the same form as equation (3.22) and represents precession of $\boldsymbol{S}$ around the magnetic field direction with precession frequency eB/m. This frequency is known as the Larmor precession frequency. It can be detected by observing the absorption of power from an applied oscillating electric field in a resonance experiment. The absorption shows a sharp peak when the frequency of the electric field is equal to the Larmor frequency. This is the basis of the very powerful experimental technique known as electron spin resonance spectroscopy, or ESR for short, which is used in many branches of experimental physics and chemistry for studying the magnetic fields in atoms, molecules and solids.

4 The Gradient Operator

4.1 Scalar and Vector Fields

In this section we will be dealing with functions of more than one variable. For instance the potential energy of a particle moving in some force field may depend on its position $\boldsymbol{r}$, i.e. it is a function of its x-, y- and z-coordinates if we are using a Cartesian system. The potential energy $V(\boldsymbol{r})$, or $V(x, y, z)$ is said to be a 'scalar field'. It has a definite, scalar, value at every point in space. Another example of a scalar field would be the temperature distribution $T(\boldsymbol{r})$ throughout a body.

In a spatially varying electric field the value of the field $\boldsymbol{E}$ at the point $\boldsymbol{r}$ will also be a function of the three variables x, y and z, but this time it has a vector value at each point in space. The electric field $\boldsymbol{E}(\boldsymbol{r})$ or $\boldsymbol{E}(x, y, z)$ is therefore described as a 'vector field'. Another example of a vector field would be the velocity of flow $\boldsymbol{v}(\boldsymbol{r})$, of a fluid.

4.2 Partial Derivatives

We will need to take derivatives of these fields and hence we briefly review the concept of partial differentiation.

Suppose we have a scalar function $f(x, y)$ depending on two variables, x and y (a scalar field). If we keep the variable y fixed and vary x the function will change and the partial derivative of $f(x, y)$ with respect to x is defined in exactly the same way as the ordinary derivative, that is

$$\frac{\partial}{\partial x} f(x, y) = \lim_{\delta x \to 0} \frac{f(x + \delta x, y) - f(x, y)}{\delta x}. \tag{4.1}$$

As an example, consider the function $f(x, y) = x^3 \cos y$. The partial derivatives with respect to x and y are

$$\frac{\partial f}{\partial x} = 3x^2 \cos y \qquad \text{and} \quad \frac{\partial f}{\partial y} = -x^3 \sin y.$$

The partial derivative of a vector field follows in exactly the same way as in §3.1, the result being another vector field.

To calculate the small change in a function $f(x, y)$ when x is changed by a small amount δx we have from equation (4.1)

$$\delta f(x, y) = f(x + \delta x, y) - f(x, y) \simeq \frac{\partial}{\partial x} f(x, y)\delta x.$$

If we have small changes in both x and y, then the change in $f(x, y)$ becomes

$$\begin{aligned}\delta f &= f(x + \delta x, y + \delta y) - f(x, y)\\ &= f(x + \delta x, y + \delta y) - f(x, y + \delta y) + f(x, y + \delta y) - f(x, y)\\ &\simeq \frac{\partial}{\partial x} f(x, y + \delta y)\delta x + \frac{\partial}{\partial y} f(x, y)\delta y.\end{aligned} \tag{4.2}$$

In the limit as δx and δy go to zero we write

$$\mathrm{d}f = \lim_{\delta x, \delta y \to 0} \delta f = \frac{\partial}{\partial x} f(x, y)\,\mathrm{d}x + \frac{\partial}{\partial y} f(x, y)\,\mathrm{d}y \tag{4.3a}$$

where $\mathrm{d}f$, $\mathrm{d}x$ and $\mathrm{d}y$ are now infinitesimally small. Clearly we can generalise to three dimensions and write

$$\mathrm{d}f = \frac{\partial f}{\partial x}\mathrm{d}x + \frac{\partial f}{\partial y}\mathrm{d}y + \frac{\partial f}{\partial z}\mathrm{d}z. \tag{4.3b}$$

This result essentially says that the change in a function of several variables is the sum of the changes produced by small displacements in each of the directions separately. (If we use equation (4.3) for finite displacements the error introduced by replacing $f(x, y + \delta y)$ by $f(x, y)$ in going from equation (4.2) to (4.3*a*) is second order in the small quantities δx, δy so that (4.3) is still a good approximation for small displacements.)

4.3 The Gradient Operator

In this section we are going to introduce the very important idea of a vector operator by considering the relationship between the work done by a force and the energy change of a system. In §2.1 we noted that the work done by a force $\boldsymbol{F}$, which moves its point of application a vector distance $\boldsymbol{d}$, is

$$W = \boldsymbol{F} \cdot \boldsymbol{d}.$$

We now consider the case of a force which varies with position (i.e. a force field) $\boldsymbol{F}(\boldsymbol{r})$. If this force moves a body through an infinitesimally small displacement $\mathrm{d}\boldsymbol{r}$, the work done by the force will be

$$\mathrm{d}W = \boldsymbol{F}(\boldsymbol{r}) \cdot \mathrm{d}\boldsymbol{r}. \tag{4.4}$$

As a result of this displacement there will be a change in the kinetic energy of the body. This can be calculated from Newton's Law

$$\begin{aligned} \mathrm{d}(\textsc{ke}) &= \mathrm{d}(\tfrac{1}{2}mv^2) = \mathrm{d}(\tfrac{1}{2}m\boldsymbol{v}\cdot\boldsymbol{v}) = \tfrac{1}{2}m(\boldsymbol{v}\cdot\mathrm{d}\boldsymbol{v} + \mathrm{d}\boldsymbol{v}\cdot\boldsymbol{v}) \\ &= m\mathrm{d}\boldsymbol{v}\cdot\boldsymbol{v} = m\frac{\mathrm{d}\boldsymbol{v}}{\mathrm{d}t}\cdot\boldsymbol{v}\mathrm{d}t = m\boldsymbol{a}\cdot\boldsymbol{v}\mathrm{d}t = m\boldsymbol{a}\cdot\mathrm{d}\boldsymbol{r} \\ &= \boldsymbol{F}\cdot\mathrm{d}\boldsymbol{r} \end{aligned}$$

i.e. the change in kinetic energy is just equal to the work done by the force. For a 'conservative' force (all the microscopic forces of physics are conservative, see discussion later) we can define a potential energy function $V(\boldsymbol{r})$ which is a scalar function of position such that the work done is just minus the change in the potential energy i.e.

$$\mathrm{d}V = -\mathrm{d}W = -\boldsymbol{F}\cdot\mathrm{d}\boldsymbol{r}. \tag{4.5}$$

In this case the total energy, defined to be the sum of the kinetic and potential energies, is conserved:

$$\mathrm{d}E = \mathrm{d}(\textsc{ke}) + \mathrm{d}V = 0. \tag{4.6}$$

This equation expresses the well known physical law of conservation of energy and is the origin of the term 'conservative' used above to describe a force which can be written in terms of a potential energy function. (We note that equation (4.5) only defines changes in the potential energy so that $V(\boldsymbol{r})$ is only defined to within an arbitrary additive constant. This constant is usually chosen so that $V(\boldsymbol{r})$ is zero at $\boldsymbol{r} = \infty$.)

The simplest example of potential energy that we usually meet is that associated with the gravitational field near the earth. Here the gravitational force is just $-mg\hat{\boldsymbol{k}}$ (where $\hat{\boldsymbol{k}}$ is a unit vector in the vertical direction). No work is done in a horizontal displacement, so the potential energy is just a function of the height z. In a vertical displacement, the work done by the gravitational field is $-mg\mathrm{d}z$ (i.e. positive for a downward displacement). Hence, from equation (4.5), the change in potential energy is

$$\mathrm{d}V = -\mathrm{d}W = mg\mathrm{d}z.$$

Integrating this equation gives

$$V = mgz + C \tag{4.7}$$

where C is an arbitrary constant which is usually chosen to be zero.

We now want to invert the relation (4.5) to give $\boldsymbol{F}(\boldsymbol{r})$ in terms of $V(\boldsymbol{r})$. To do this we first rewrite (4.5) in component form

$$\mathrm{d}V = -F_x\mathrm{d}x - F_y\mathrm{d}y - F_z\mathrm{d}z$$

and compare this equation with equation (4.3) with $f = V$ i.e.

$$\mathrm{d}V = \frac{\partial V}{\partial x}\mathrm{d}x + \frac{\partial V}{\partial y}\mathrm{d}y + \frac{\partial V}{\partial z}\mathrm{d}z.$$

From these two equations we see that we can write

$$F_x = -\frac{\partial V}{\partial x} \qquad F_y = -\frac{\partial V}{\partial y} \qquad F_z = -\frac{\partial V}{\partial z}.$$

This set of equations can clearly be written as a single vector equation

$$\boldsymbol{F} = -\hat{\boldsymbol{i}}\frac{\partial V}{\partial x} - \hat{\boldsymbol{j}}\frac{\partial V}{\partial y} - \hat{\boldsymbol{k}}\frac{\partial V}{\partial z} \tag{4.8}$$

and it is usual to write this in the form

$$\boldsymbol{F} = -\text{grad}V \qquad \text{or } \boldsymbol{F} = -\nabla V \tag{4.9}$$

where the vector grad$V$ or ∇V is defined by equation (4.8) to be

$$\text{grad}V = \nabla V = \hat{\boldsymbol{i}}\frac{\partial V}{\partial x} + \hat{\boldsymbol{j}}\frac{\partial V}{\partial y} + \hat{\boldsymbol{k}}\frac{\partial V}{\partial z}. \tag{4.10}$$

We now go one step further and think of the operator ∇ as a vector operator with x-component $\partial/\partial x$ etc, i.e.

$$\nabla = \hat{\boldsymbol{i}}\frac{\partial}{\partial x} + \hat{\boldsymbol{j}}\frac{\partial}{\partial y} + \hat{\boldsymbol{k}}\frac{\partial}{\partial z}. \tag{4.11}$$

The gradient operator ∇ is pronounced 'del'. We do not wish to go into detail here but to describe an operator as a vector strictly means that it must transform in the same way as an ordinary vector under a rotation of the coordinate axes. It can be shown that this is the case for the operator ∇. From our point of view it is really equation (4.9) or (4.10) which is significant, i.e. if we operate with ∇ on any scalar function such as $V(\boldsymbol{r})$ we get a vector field.

Although we have defined the gradient operator via equation (4.11) in terms of a particular coordinate system, it is important that equations such as (4.9) have a meaning independent of a coordinate system. To get some insight into this meaning we will look at a particular example where we are familiar with the results. We consider a particle sliding without friction on the slopes of a hill and ask the question—'what is the magnitude and direction of the resulting force on the particle at any point on the hillside?' Let us assume that we can describe the shape of the hill by a contour map as in figure 4.1. The contour lines link together points of equal height z. The gravitational potential energy of the particle at any point on the hill will be given by equation (4.7) as $V(x, y) = mgz(x, y)$, where $z(x, y)$ is the height at the point (x, y). The force on the particle will be given by equation (4.9), $\boldsymbol{F} = -\nabla V$. ∇V is thus a vector with magnitude equal to that of the force but pointing in the opposite direction.

Consider the particle at some point such as P on one of the contours in figure 4.1. Now let it make a small displacement d$\boldsymbol{r}$ from P. The change in potential energy will be given by

$$\mathrm{d}V = -\boldsymbol{F}\cdot\mathrm{d}\boldsymbol{r} = \nabla \boldsymbol{V}\cdot\mathrm{d}\boldsymbol{r}. \tag{4.12}$$

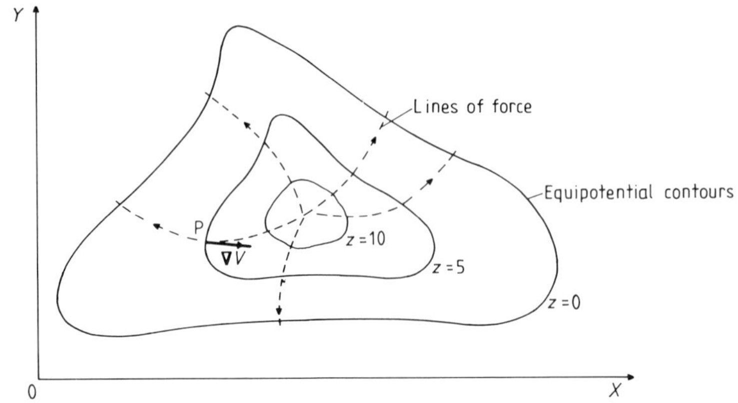

Figure 4.1

Now, if $d\boldsymbol{r}$ is chosen to be along the contour, dV will be zero so that $\nabla V \cdot d\boldsymbol{r}$ must also be zero. Hence the vector ∇V must be perpendicular to $d\boldsymbol{r}$ i.e. perpendicular to the contour of constant V. It also follows from equation (4.12) that if the displacement $d\boldsymbol{r}$ is of length ds and makes an angle θ with the direction of ∇V the change in V will be $\nabla V ds \cos\theta$. The change dV is thus a maximum when the displacement is in the direction of ∇V, and in this case $|\nabla V| = dV/ds$. Equation (4.12) thus gives us a good picture of ∇V. It is a vector pointing in the direction of the maximum slope at P (i.e. normal to the contour of constant V) with size equal to the maximum rate of change of V at P i.e. equal to the rate of change of V in the direction of maximum slope.

Although we introduced equation (4.12) by considering the gravitational force on a particle we could have written it down directly from equation (4.3) which is simply

$$df = \nabla f \cdot d\boldsymbol{r} \tag{4.13}$$

expanded in component form. The conclusions we drew are therefore quite general and we can regard (4.13) as a defining equation for the gradient operator in a form which is independent of the choice of coordinates.

The example we have been considering had a scalar potential $V(x, y)$ which was a function of two variables x and y. In general we meet functions of three (or more) variables. The contours of constant V which we met above then become surfaces (or hypersurfaces) of constant V, and $\nabla V(\boldsymbol{r})$ is a vector normal to the surface at $\boldsymbol{r}$ with magnitude equal to the rate of change of $V(\boldsymbol{r})$ in the direction of this normal.

4.4 Examples of the Gradient Operator

Example 4.1 Temperature gradient in two dimensions
Let $f(\boldsymbol{r})$ be the scalar field representing the temperature at the point $\boldsymbol{r}$ in the x–y plane. We choose

$$f(x, y) = T(x, y) = -x^2 + y \tag{4.14}$$

so that the 'surfaces' of constant f (constant temperature) in this case will be the parabolas

$$y = x^2 + T$$

We illustrate a set of these surfaces for $T = -1$, $T = 0$, $T = 1$ and $T = 2$, in figure 4.2.

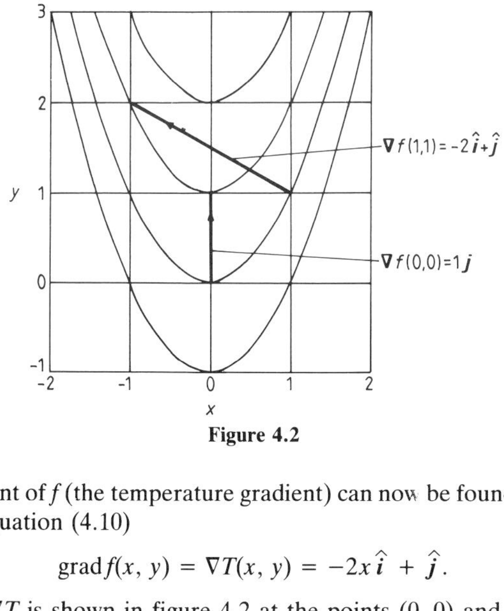

Figure 4.2

The gradient of f (the temperature gradient) can now be found at any point x, y using equation (4.10)

$$\text{grad} f(x, y) = \nabla T(x, y) = -2x\hat{\imath} + \hat{\jmath}.$$

The vector ∇T is shown in figure 4.2 at the points (0, 0) and (1, 1) on the contour $T = 0$.

Example 4.2 Spherically symmetric fields
We frequently meet fields in physics which are spherically symmetric so that $f(\boldsymbol{r})$ is a function only of the radial distance from some origin i.e. $f(r)$. Examples 4.3 and 4.4 which follow are of this type so we first calculate the gradient of an arbitrary scalar function. Although it would seem obvious that it would be better to use spherical polar coordinates to do this, we want to avoid introducing them at this stage and the calculation is quite straightforward in the Cartesian form for which we have the gradient operator. Since

$r = (x^2 + y^2 + z^2)^{1/2}$ we have

$$\frac{\partial f(r)}{\partial x} = \frac{\partial r}{\partial x}\frac{\mathrm{d}f(r)}{\mathrm{d}r} = \tfrac{1}{2}(x^2 + y^2 + z^2)^{-1/2}2xf'$$
$$= (x/r)f'$$

where f' is the derivative $\mathrm{d}f(r)/\mathrm{d}r$. Hence

$$\nabla f(r) = \left(\frac{x}{r}\hat{\boldsymbol{i}} + \frac{y}{r}\hat{\boldsymbol{j}} + \frac{z}{r}\hat{\boldsymbol{k}}\right)f'(r)$$
$$= \hat{\boldsymbol{r}}f'(r) \qquad \text{or} = f'(r)\hat{\boldsymbol{r}}. \tag{4.15}$$

Example 4.3 Gravitational fields

The conservative gravitational and electrostatic force fields of nature can be writen as gradients of appropriate scalar potential fields. For instance, we can write the gravitational force field due to a mass m at the origin:

$$\boldsymbol{g} = -(Gm/r^2)\hat{\boldsymbol{r}} \qquad \text{or } \boldsymbol{g} = -(Gm/r^3)\boldsymbol{r} \tag{4.16}$$

in terms of the gradient of the gravitational potential, $\boldsymbol{g} = -\nabla\varphi$, where

$$\varphi = Gm/r. \tag{4.17}$$

This result follows immediately from equation (4.15) since φ is a spherically symmetric potential.

In this example, the equipotential surfaces are spheres centred at the origin and the gravitational field lines are radial (i.e. normal to these surfaces).

Example 4.4 Electrostatic fields

The electrostatic field has exactly the same form as the gravitational field so that, in analogy with equations (4.16) and (4.17), we can write the fields due to a point charge q at the origin

$$\boldsymbol{E} = (kq/r^2)\hat{\boldsymbol{r}} \qquad \text{or } \boldsymbol{E} = (kq/r^3)\boldsymbol{r} \tag{4.18}$$

and $\boldsymbol{E} = -\nabla\varphi$, where

$$\varphi = -kq/r \tag{4.19}$$

and the constant k is usually written as $1/4\pi\varepsilon_0$, in terms of the permittivity of free space ε_0.

Example 4.5 Electrical conductivity

The electrical current density $\boldsymbol{j}(\boldsymbol{r})$ in a medium of conductivity σ is given in terms of the electrical potential φ by

$$\boldsymbol{j}(\boldsymbol{r}) = -\nabla\varphi(\boldsymbol{r}). \tag{4.20}$$

This is essentially Ohm's law as we can see by applying it to a rod with a cross sectional area A and length L carrying a total current i. In this case the current density j is i/A and if the potential φ drops by V volts from one end of the rod to the other $\nabla\varphi = -V/L$. Equation (4.20) thus becomes

$$i/A = \sigma V/L. \tag{4.21}$$

The resistance R of the rod is $\rho L/A$ where the resistivity ρ is $1/\sigma$, i.e. $\sigma = L/RA$. Substituting this in (4.21) gives Ohm's law $V = iR$.

Example 4.6 Heat conductivity
When heat flows in a medium (with no internal sources of heat) it is found experimentally that the heat current flows down the temperature gradient and can be described by the equation

$$\boldsymbol{j}(\boldsymbol{r}) = -K\nabla\theta(\boldsymbol{r}). \tag{4.22}$$

Here $\theta(\boldsymbol{r})$ is the temperature at the point $\boldsymbol{r}$, $\boldsymbol{j}(\boldsymbol{r})$ is the heat current at $\boldsymbol{r}$ (i.e. the amount of heat flowing across unit area of a surface perpendicular to the direction of $\boldsymbol{j}(\boldsymbol{r})$, per second) and K is a constant called the thermal conductivity.

Example 4.7 Diffusion
This is similar to heat flow but in this case it is molecules that are flowing.

Consider a gas with a density of molecules $n(\boldsymbol{r})$ which varies with position. The molecules move to try to even out the density and again, experimentally, the current of particles $\boldsymbol{j}(\boldsymbol{r})$ is found to satisfy an equation (Fick's law)

$$\boldsymbol{j}(\boldsymbol{r}) = -D\nabla n(\boldsymbol{r}) \tag{4.23}$$

where D is the diffusion coefficient.

The essential similarity of the three physical processes discussed in these last three examples is clearly demonstrated by the form of the equations (4.20), (4.22) and (4.23).

The Divergence of a Vector Field

5

In chapter 4 we saw that the differential operator ∇ could be regarded as a vector and in this chapter and the next we consider the results of taking the scalar and vector products of this operator with a vector field.

If we take the scalar product of the vector operator ∇ with a vector field $\boldsymbol{F}(\boldsymbol{r})$ we will produce a scalar quantity $\nabla\cdot\boldsymbol{F}(\boldsymbol{r})$ at every point $\boldsymbol{r}$, i.e. a scalar field $f(r) = \nabla\cdot\boldsymbol{F}(\boldsymbol{r})$.

In Cartesian coordinates this becomes

$$\begin{aligned}\nabla\cdot\boldsymbol{F}(\boldsymbol{r}) &= \left(\hat{\boldsymbol{i}}\frac{\partial}{\partial x} + \hat{\boldsymbol{j}}\frac{\partial}{\partial y} + \hat{\boldsymbol{k}}\frac{\partial}{\partial z}\right)\cdot(F_x\hat{\boldsymbol{i}} + F_y\hat{\boldsymbol{j}} + F_z\hat{\boldsymbol{k}}) \\ &= \frac{\partial F_x}{\partial x} + \frac{\partial F_y}{\partial y} + \frac{\partial F_z}{\partial z}.\end{aligned} \tag{5.1}$$

The scalar field $\nabla\cdot\boldsymbol{F}$ is called the divergence of $\boldsymbol{F}$ and is often written div $\boldsymbol{F}$. The physical significance of the divergence is not at all clear from equation (5.1), its component form, so we move on to consider situations where the divergence arises naturally to get a better feeling for its meaning. Unfortunately before we can do this it is necessary to digress for a few pages to look at line, surface and volume integrals which will occur frequently in the remaining sections.

5.1 Line, Surface and Volume Integrals

5.1.1 *Line Integrals*

Let us consider the work done on a particle which travels along a definite path in a force field $\boldsymbol{F}(\boldsymbol{r})$. The work done by the field in any infinitesimal displacement $\mathrm{d}\boldsymbol{r}$ will be given as in equation (4.4) by

$$\mathrm{d}W = \boldsymbol{F}(\boldsymbol{r})\cdot\mathrm{d}\boldsymbol{r}.$$

The total work done in following a particular curve C from an initial position i to a final position f will be the sum of contributions $\mathrm{d}W_i$ from each section

$\mathrm{d}\boldsymbol{r}_i$ in the limit that the sections $\mathrm{d}\boldsymbol{r}_i$ tend to zero i.e.

$$W = \lim_{\mathrm{d}r_i \to 0} \sum_i \boldsymbol{F}(\boldsymbol{r}_i) \cdot \mathrm{d}\boldsymbol{r}_i.$$

This is illustrated in figure 5.1.

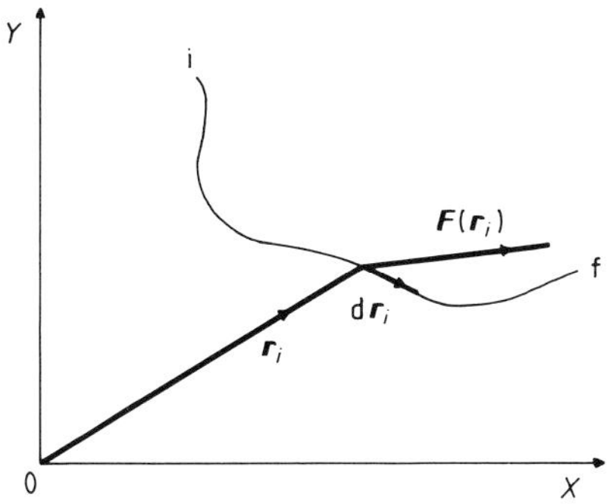

Figure 5.1

This limit is called the line integral of $\boldsymbol{F}(\boldsymbol{r})$ along the curve C and is written as

$$W = \lim_{\mathrm{d}r_i \to 0} \sum_i \boldsymbol{F}(\boldsymbol{r}_i) \cdot \mathrm{d}\boldsymbol{r}_i = \int_C \boldsymbol{F}(\boldsymbol{r}) \cdot \mathrm{d}\boldsymbol{r}. \tag{5.2}$$

Line integrals of the type in equation (5.2) arise frequently in physics. The evaluation of the line integral for general curves can be quite difficult and it is usually necessary to go back to Cartesian (or other) coordinates to do this. We will look at an example below but, for most physically important situations the geometry of the curve and vector field will be such that the integral can be easily carried out—often by inspection! For this reason we will not go into detail here.

As a particularly simple example the work done by a conservative force which we can write as $\boldsymbol{F} = -\nabla V$, in terms of a scalar potential V, will be

$$W = -\int_C \nabla V \cdot \mathrm{d}\boldsymbol{r} = -\int_C \mathrm{d}V$$

where we have used equation (4.12). The line integral is now just minus the sum of the changes in V as we follow the line and is simply the difference between the initial and final values of V so that

$$W = V_\mathrm{i} - V_\mathrm{f}.$$

In this special case the line integral has a value which is independent of the shape of the line. Thus the work done by a conservative force in moving a particle from one point to another is independent of the path taken. This is in fact an alternative way of defining a conservative force. For a non-conservative force, such as friction, the work done would be path dependent—more work would be done on longer paths!

If the path of integration forms a closed curve so that the initial and final points coincide the integral is usually written in the form

$$\oint \boldsymbol{F}(\boldsymbol{r})\cdot \mathrm{d}\boldsymbol{r}.$$

For a conservative force field this line integral round a closed loop will always be zero since the potential is the same at the initial and final points which coincide.

As an example of a line integral which does depend on the path we consider a vector field $\boldsymbol{F}(\boldsymbol{r}) = x^2\hat{\boldsymbol{i}} + xy\hat{\boldsymbol{j}}$ and consider two paths for line integrals from the origin (0, 0) to the point (1, 1). These paths are illustrated in figure 5.2.

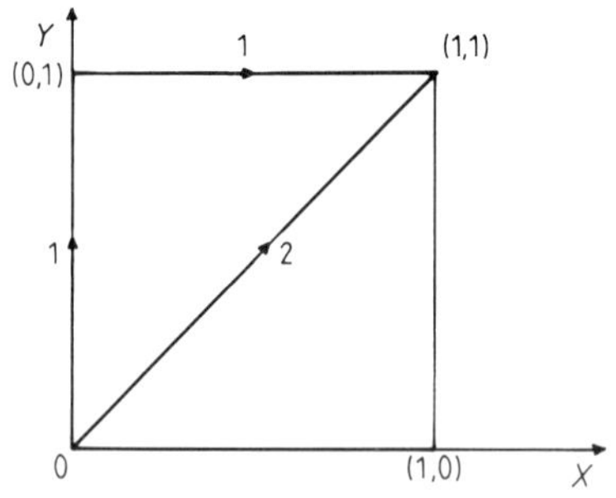

Figure 5.2

The first path goes up the y-axis to the point (0, 1) and then parallel to the x-axis to (1, 1). For this path we have

$$I_1 = \int_{C_1} \boldsymbol{F}\cdot \mathrm{d}\boldsymbol{r} = \int_{C_1} F_x\,\mathrm{d}x + \int_{C_1} F_y\,\mathrm{d}y = \underset{(\text{with } x=0)}{\int_0^1 F_y\,\mathrm{d}y} + \underset{(\text{with } y=1)}{\int_0^1 F_x\,\mathrm{d}x}$$

since $\mathrm{d}x$ will be zero for any element up the y-axis and $\mathrm{d}y$ will be zero for any element parallel to the x-axis. Hence

$$I_1 = \underset{(\text{with } x=0)}{\int_0^1 xy\,\mathrm{d}y} + \underset{(\text{with } y=1)}{\int_0^1 x^2\,\mathrm{d}x} = 0 + \tfrac{1}{3}x^3 = \tfrac{1}{3}.$$

The second path goes directly from the origin to the point (1, 1) along the straight line $x = y$. For this path we have

$$I_2 = \int_0^1 x^2\,\mathrm{d}x + \underset{(\text{with } y=x)}{\int_0^1 xy\,\mathrm{d}y} = \int_0^1 x^2\,\mathrm{d}x + \int_0^1 y^2\,\mathrm{d}y = \tfrac{2}{3}$$

where we have used the fact that $x = y$ in the second integral. Thus the two results are different for this non-conservative force field. (You may like to check that it is impossible to write this field as the gradient of a scalar potential field.)

Line integrals can also be defined in exactly the same way for scalar fields e.g.

$$\boldsymbol{L} = \int V(\boldsymbol{r})\,\mathrm{d}\boldsymbol{r} = \lim_{\mathrm{d}\boldsymbol{r}_i \to 0} \sum_i V(\boldsymbol{r}_i)\,\mathrm{d}\boldsymbol{r}_i. \tag{5.3}$$

$\boldsymbol{L}$ is a vector quantity in this case since each contribution to the sum in (5.3) is a vector.

We can also form a vector from the line integral

$$\int \boldsymbol{F}(\boldsymbol{r}) \wedge \mathrm{d}\boldsymbol{r} = \lim_{\mathrm{d}\boldsymbol{r}_i \to 0} \sum_i \boldsymbol{F}(\boldsymbol{r}_i) \wedge \mathrm{d}\boldsymbol{r}_i \tag{5.4}$$

although this rarely occurs in physical problems.

5.1.2 *Surface Integrals*

In the same way that we defined line integrals in the previous section we can define integrals of scalar or vector fields over surfaces. It is first necessary to characterise a small element of surface area.

If the surface is flat then we use a vector $\mathrm{d}\boldsymbol{s} = \mathrm{d}s\,\hat{\boldsymbol{n}}$ to represent a small surface element of area $\mathrm{d}s$ where $\hat{\boldsymbol{n}}$ is a unit vector normal to the surface as in figure 5.3. There is clearly a small problem here in choosing a sense for $\hat{\boldsymbol{n}}$ and for the moment we leave this choice arbitrary.

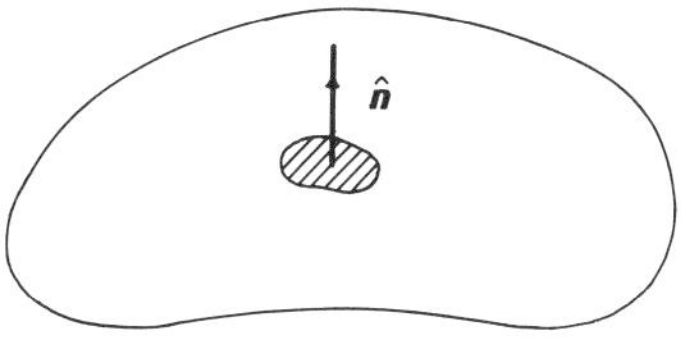

Figure 5.3

For a curved surface we can still use the same definition in the limit as $\mathrm{d}s$ goes to zero when the element $\mathrm{d}\boldsymbol{s} = \mathrm{d}s\,\hat{\boldsymbol{n}}$ is still well defined.

In the case of a closed surface in three dimensions (the case we will meet most frequently) the sense of $\hat{\boldsymbol{n}}$ is conventionally chosen as the outward direction.

We can now define surface integrals as in the previous section, as the limit of a sum—i.e. the scalar surface integral of a vector field $\boldsymbol{F}(\boldsymbol{r})$ over a surface S is

$$\int_S \boldsymbol{F}(\boldsymbol{r})\cdot\mathrm{d}\boldsymbol{s} = \lim_{\mathrm{d}\boldsymbol{r}_i \to 0} \sum_i \boldsymbol{F}(\boldsymbol{r}_i)\cdot\mathrm{d}\boldsymbol{s}_i. \tag{5.5}$$

The surface integral defined in equation (5.5) is also called the flux of the vector field $\boldsymbol{F}(\boldsymbol{r})$ through the surface $\boldsymbol{S}$. We can understand this interpretation if we think of $\boldsymbol{F}(\boldsymbol{r})$ as a vector current (of particles for instance), usually written $\boldsymbol{j}(\boldsymbol{r})$. In this case the direction of $\boldsymbol{j}(\boldsymbol{r})$ represents the direction of particle motion at the point $\boldsymbol{r}$ and the magnitude of $\boldsymbol{j}(\boldsymbol{r})$ represents the number of particles per second flowing across unit surface area of the plane perpendicular to $\boldsymbol{j}(\boldsymbol{r})$. The quantity $\boldsymbol{j}(\boldsymbol{r})\cdot\mathrm{d}\boldsymbol{s}$ then represents the number of particles flowing across the surface $\mathrm{d}\boldsymbol{s}$ per second, and the surface integral is the total (net) number of particles crossing the area $\boldsymbol{S}$ per second, which is the usual meaning of the word flux.

In the case where the surface is closed, the integral is usually written

$$\oint_S \boldsymbol{F}(\boldsymbol{r})\cdot\mathrm{d}\boldsymbol{s} \tag{5.6}$$

and clearly represents the flux of $\boldsymbol{F}(\boldsymbol{r})$ out of the closed region of space defined by the surface $\boldsymbol{S}$. This will have important physical applications. For instance, if $\boldsymbol{j}(\boldsymbol{r})$ represents a current of particles, the flux out of some region must equal the rate of loss of particles from the region. We use this result in §5.4 to derive an equation of continuity for $\boldsymbol{j}(\boldsymbol{r})$.

The actual evaluation of surface integrals, in general cases, is difficult, but, as for line integrals, the cases we meet in situations of physical importance often have high symmetry making the evaluation trivial. For instance, if we have a spherically symmetric field such as the electric field due to a point charge at the origin, and a surface integral over a sphere of radius a centred at the origin, the field is always perpendicular to the surface and hence parallel to the normal $\hat{\boldsymbol{n}}$. It will also be constant over the surface so that we get the simple result

$$\begin{aligned}\oint \boldsymbol{E}(\boldsymbol{r})\cdot\mathrm{d}\boldsymbol{s} &= \oint E(a)\hat{\boldsymbol{r}}\cdot\hat{\boldsymbol{r}}\;\mathrm{d}s = E(a)\oint \mathrm{d}s \\ &= 4\pi a^2 E(a) = q/\varepsilon_0 \end{aligned} \tag{5.7}$$

where we have used the expressión (4.18) for the electric field. Notice that this result for the flux of $\boldsymbol{E}$ though a sphere of radius a is independent of the radius a.

5.1.3 *Volume Integrals*

We characterise a small volume element by the symbol $\mathrm{d}V_i$ and this is illustrated in figure 5.4.

In this case the element is a scalar quantity as opposed to the vector element for line and surface integrals. By dividing a region of space into a large number of elements $\mathrm{d}V_i$ we can define the volume integral of a vector field $\boldsymbol{F}(\boldsymbol{r})$ over the region V in the now familiar way as

$$\boldsymbol{I} = \int_V \boldsymbol{F}(\boldsymbol{r})\,\mathrm{d}V = \lim_{\mathrm{d}V_i\to 0}\sum_i \boldsymbol{F}(\boldsymbol{r}_i)\,\mathrm{d}V_i. \tag{5.8}$$

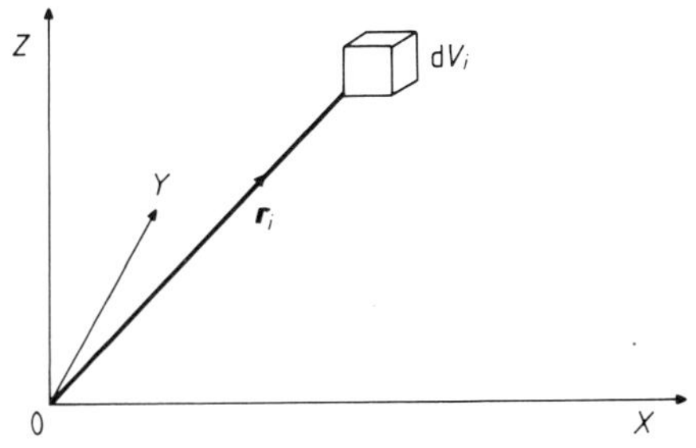

Figure 5.4

A similar result holds for a scalar field. Again evaluation of these integrals for arbitrary fields and regions of space is very difficult but in physically interesting situations we will usually be saved by the symmetry of the problem.

5.2 The Meaning of the Divergence of a Vector Field

Using the idea of the surface integral discussed in §5.1.2 we are now going to calculate the flux of a vector field $\boldsymbol{F}(\boldsymbol{r})$ out of an infinitesimally small volume around the point $\boldsymbol{r}$. We show that the divergence of $\boldsymbol{F}(\boldsymbol{r})$ at the point $\boldsymbol{r}$, defined in equation (5.1), is equal to the value of this flux divided by the infinitesimal volume, i.e.

$$\operatorname{div}\boldsymbol{F}(\boldsymbol{r}) = \lim_{\mathrm{d}V\to 0}\frac{\text{flux of } \boldsymbol{F}(\boldsymbol{r}) \text{ out of } \mathrm{d}V}{\mathrm{d}V} \tag{5.9}$$

i.e.

$$\operatorname{div}\boldsymbol{F}(\boldsymbol{r}) = \lim_{\mathrm{d}V\to 0}\frac{1}{\mathrm{d}V}\oint \boldsymbol{F}(\boldsymbol{r})\cdot \mathrm{d}\boldsymbol{s}. \tag{5.10}$$

To do this we evaluate the right-hand side of this equation when the small volume is a cube with sides from x to $x + \mathrm{d}x$, y to $y + \mathrm{d}y$ and z to $z + \mathrm{d}z$, as in figure 5.5.

The integral must be taken over all six faces of the small cube. If we first look at the contribution to the integral from the two faces perpendicular to the x-direction then $\mathrm{d}\boldsymbol{s}$ is in the direction $-x$ for the left-hand face and $+x$ for the right-hand face so the only component of $\boldsymbol{F}$ which contributes is the x-component. This gives the contribution

$$\frac{1}{\mathrm{d}V}(-F_x(x, y, z)\mathrm{d}y\mathrm{d}z + F_x(x + \mathrm{d}x, y, z)\mathrm{d}y\mathrm{d}z.$$

We now write $F_x(x + \mathrm{d}x, y, z)$ as $F_x(x, y, z) + (\partial F_x(x, y, z)/\partial x)\mathrm{d}x$ and use the fact that $\mathrm{d}V$ is $\mathrm{d}x\mathrm{d}y\mathrm{d}z$ to reduce this to the expression

$$\frac{1}{\mathrm{d}x\mathrm{d}y\mathrm{d}z}\frac{\partial}{\partial x}F_x(x, y, z)\mathrm{d}x\mathrm{d}y\mathrm{d}z = \frac{\partial F_x}{\partial x}.$$

In a similar way the contributions from the other pairs of faces give $\partial F_y/\partial y$ and $\partial F_z/\partial z$. We thus have

$$\begin{aligned}\frac{1}{\mathrm{d}V}\oint \boldsymbol{F}(\boldsymbol{r})\cdot\mathrm{d}\boldsymbol{s} &= \frac{\partial F_x}{\partial x} + \frac{\partial F_y}{\partial y} + \frac{\partial F_z}{\partial z} \\ &= \nabla\cdot\boldsymbol{F} \end{aligned} \qquad (5.11)$$

which is our original form for the divergence of $\boldsymbol{F}$ (equation (5.1)). (We have not attempted to be rigorous in deriving this result. For instance we have ignored the fact that $\boldsymbol{F}_x$ changes over the face while keeping the change between faces. This can be justified and the result can be shown to be independent of the choice of shape for the volume element.)

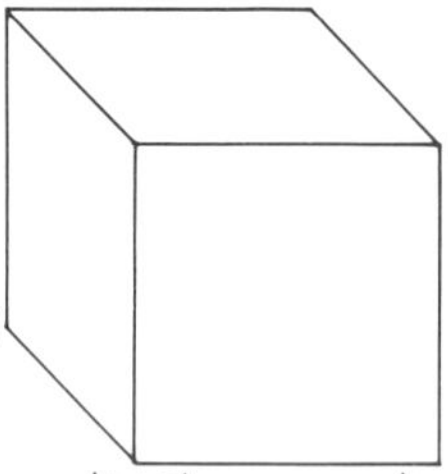

Figure 5.5

The expression (5.9), or equation (5.10), contains the essential physical nature of the divergence of $\boldsymbol{F}$ as a flux of $\boldsymbol{F}$ out of some region. It will only be non-zero if there is a source (or sink) of $\boldsymbol{F}$ in that region. By considering the case where $\boldsymbol{F}(\boldsymbol{r})$ is a current of particles we can see that $\operatorname{div}\boldsymbol{F}(\boldsymbol{r})$ must represent the number of particles created per second in the volume $\mathrm{d}V$ divided by $\mathrm{d}V$ i.e. the source density at $\boldsymbol{r}$. If $\boldsymbol{F}(\boldsymbol{r})$ is the heat current, then $\operatorname{div}\boldsymbol{F}(\boldsymbol{r})$ is the amount of heat generated per second, per unit volume, at $\boldsymbol{r}$.

5.3 Gauss's Theorem

This theorem, which is also known as the divergence theorem, can be written

$$\int \operatorname{div}\boldsymbol{F}\,\mathrm{d}V = \oint \boldsymbol{F}\cdot\mathrm{d}\boldsymbol{s} \qquad (5.12)$$

where the surface integral on the right is over the bounding surface of the volume integral on the left. It follows directly from the form of the divergence given in equation (5.10) if we divide this volume up into a large number of small volume elements $\mathrm{d}V_i$ as in figure 5.6.

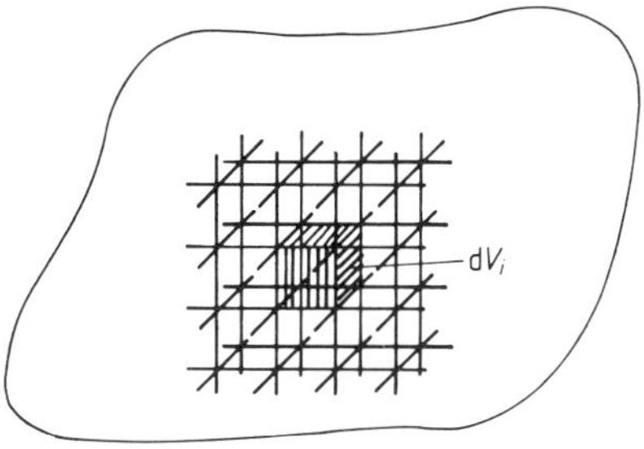

Figure 5.6

For each volume element we have from equation (5.10)

$$\operatorname{div} \boldsymbol{F}\, \mathrm{d}V_i = \oint \boldsymbol{F} \cdot \mathrm{d}\boldsymbol{s}_i$$

in the limit as $\mathrm{d}V_i$ goes to zero. We now sum this equation over all elements $\mathrm{d}V_i$ and take the limit. The left-hand side becomes $\int \operatorname{div} \boldsymbol{F}\, \mathrm{d}V$. On the right-hand side, the surface integrals over the common surface dividing two adjacent elements cancel since the normal points in the opposite directions in the two elements. Thus the only surface elements which do not give a vanishing contribution are those on the boundary. We thus get the result (5.12).

5.3.1 *Gauss's Law in Electrostatics*

We now use equation (5.12) to derive a very important result for electrostatic fields. In equation (5.7) we showed that the total flux of the electric field due to a point charge q through any spherical surface surrounding q is just q/ε_0, whatever the radius of the sphere. This shows that the flux is conserved i.e. no flux is generated between two spheres of radius r_1 and r_2. (It is straightforward to show directly that, when $\boldsymbol{E}$ is the field due to a point charge, $\operatorname{div} \boldsymbol{E} = 0$ for any point except the origin.) It follows from this that the flux through any closed surface surrounding q will also be q/ε_0. If we have several charges q_i then the total flux through any closed surface $\boldsymbol{S}$ will thus be given by

$$\int_S \boldsymbol{E} \cdot \mathrm{d}s = \sum_i \frac{q_i}{\varepsilon_0} \tag{5.13}$$

where the sum on the right is over all charges q_i contained within the surface S. This result is known as Gauss's law in electrostatics. It enables us to draw 'field lines' to represent electric fields. The direction of the field line follows the direction of $\boldsymbol{E}(\boldsymbol{r})$ and each line represents a unit of flux. Field lines thus start on positive charges and terminate on negative charges, and are continuous in any region where there is no charge density. Figure 5.7 illustrates the field lines for a pair of point charges q and $-q$.

The density of field lines represents the strength of the electric field.

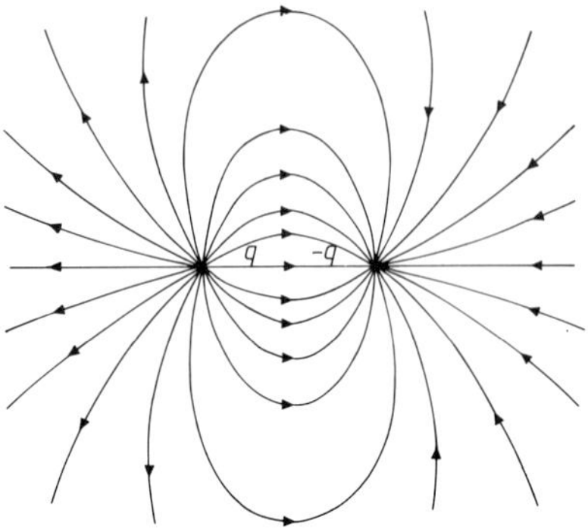

Figure 5.7

If we now consider the case of a continuous charge distribution $\rho(\boldsymbol{r})$ rather than a set of discrete charges q_i, equation (5.13) becomes

$$\oint \boldsymbol{E} \cdot \mathrm{d}\boldsymbol{s} = \int \frac{\rho \mathrm{d}V}{\varepsilon_0}.$$

We rewrite the left-hand side using Gauss's theorem (5.12) to get

$$\int \operatorname{div} \boldsymbol{E} \, \mathrm{d}V = \int \frac{\rho \mathrm{d}V}{\varepsilon_0}$$

or

$$\int \left(\operatorname{div} \boldsymbol{E} - \frac{\rho}{\varepsilon_0} \right) \mathrm{d}V = 0.$$

This must hold for any arbitrary volume so the integrand itself must vanish i.e.

$$\operatorname{div} \boldsymbol{E} = \rho / \varepsilon_0. \tag{5.14}$$

This is one of Maxwell's famous equations for electromagnetic fields.

A similar result holds for gravitational fields which have the same mathematical form, as pointed out in §4.4, i.e.

$$\int_S \boldsymbol{g} \cdot \mathrm{d}\boldsymbol{s} = -4\pi M G \tag{5.15}$$

where M is the total mass contained within the surface $\boldsymbol{S}$, and

$$\operatorname{div} \boldsymbol{g} = -4\pi \rho G. \tag{5.16}$$

In the case of magnetic fields $\boldsymbol{B}$ there are no magnetic monopoles to correspond to point charges so that we have

$$\operatorname{div} \boldsymbol{B} = 0. \tag{5.17}$$

5.4 An Equation of Continuity

In the discussion following equation (5.6), we saw that the flux of particles out of a closed region must be equal to the rate of loss of the number, N, of particles contained, assuming that the particles are conserved i.e.

$$\oint \boldsymbol{j}(\boldsymbol{r})\cdot \mathrm{d}\boldsymbol{s} = -\frac{\partial N}{\partial t}. \tag{5.18}$$

The number of particles contained will be given in terms of the density of particles $\rho(\boldsymbol{r})$ by the volume integral

$$N = \int \rho(\boldsymbol{r})\,\mathrm{d}V$$

so that equation (5.18) becomes

$$\oint \boldsymbol{j}(\boldsymbol{r})\cdot \mathrm{d}\boldsymbol{s} = -\frac{\partial}{\partial t}\int \rho(\boldsymbol{r})\,\mathrm{d}V = -\int \frac{\partial}{\partial t}\rho(\boldsymbol{r})\,\mathrm{d}V. \tag{5.19}$$

We now use Gauss's theorem (5.12) to write this as

$$\int \operatorname{div}\boldsymbol{j}(\boldsymbol{r})\,\mathrm{d}V = -\int \frac{\partial}{\partial t}\rho(\boldsymbol{r})\,\mathrm{d}V$$

or

$$\int \left(\operatorname{div}\boldsymbol{j}(\boldsymbol{r}) + \frac{\partial \rho(\boldsymbol{r})}{\partial t}\right)\mathrm{d}V = 0.$$

Since this equation holds for any volume it again follows that the integrand itself must vanish i.e.

$$\operatorname{div}\boldsymbol{j}(\boldsymbol{r}) + \frac{\partial \rho(\boldsymbol{r})}{\partial t} = 0. \tag{5.20}$$

This is called 'the equation of continuity' and holds for any conserved quanity. For instance, in electromagnetic theory $\boldsymbol{j}$ would be the electric current and ρ the charge density. In thermal physics $\boldsymbol{j}$ would be the heat current and ρ the thermal energy density $mc\theta$, where m is the mass density, c the specific heat and θ the temperature. In this case it is possible that there are heat sources in the medium and a term $\dot{Q}(\boldsymbol{r})$ would have to be added to the right-hand side of equation (5.19) to represent the rate of production of heat at that point.

For a general fluid of density ρ the current would be $\boldsymbol{j} = \rho\boldsymbol{v}$ and the equation of continuity is

$$\operatorname{div}(\rho\boldsymbol{v}) + \frac{\partial \rho}{\partial t} = \dot{q}$$

where $\dot{q}(\boldsymbol{r})$ represents a source density at the point $\boldsymbol{r}$ producing a mass $\dot{q}$ per unit volume per second.

5.5 Some Examples

(a) Field due to a spherically symmetric mass or charge distribution
First we show that the gravitational field outside a large spherically symmetric mass is the same as if all the mass were concentrated at the centre. Consider a thin spherical shell of matter of density $\rho(r)$ at radius r from the centre as in figure 5.8.

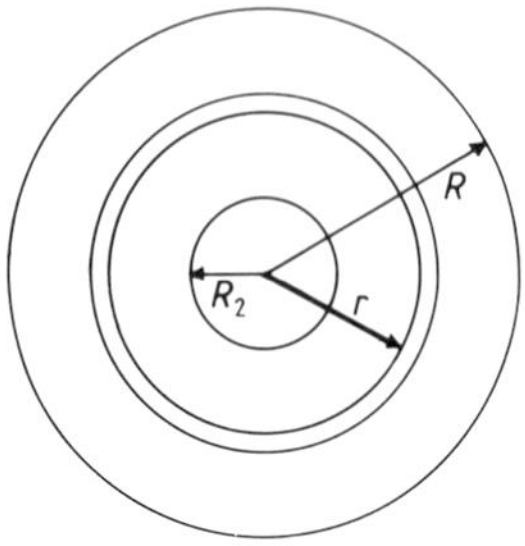

Figure 5.8

Apply Gauss's law (5.15) to the gravitational field over a spherical surface of radius R, greater than r:

$$\oint \boldsymbol{g}\cdot \mathrm{d}\boldsymbol{s} = -4\pi\Delta MG$$

where ΔM is the mass of the shell. By symmetry the field will be radial $\boldsymbol{g}(r) = g(r)\hat{\boldsymbol{r}}$ and hence normal to the surface and parallel to $\hat{\boldsymbol{n}}$. The integral can thus be carried out directly (as in equation (5.7)) to give

$$4\pi R^2 g(R) = -4\pi\Delta MG$$

i.e.

$$g(R) = -\Delta M(G/R^2)$$

which is the same as the field due to the same mass at the origin. Clearly if this is true for any spherical shell it will also be true for any spherically symmetric body of radius less than R.

We now consider the field at any point inside the spherical shell i.e. we draw a sphere of radius $R_2 < r$. In this case there is no mass contained within the sphere of radius R_2 and Gauss's law yields $g(R_2) = 0$.

For a point at radius r' inside a large spherical body we get a field which is due to the matter of radius less than r' and no field for the matter of radius greater than r', i.e.

$$g(r') = -\int_0^{r'} \frac{4\pi r^2 \rho(r)\mathrm{d}rG}{(r')^2}.$$

Clearly, analogous results will hold for the electric field due to a spherically symmetric distribution of charge if we replace the gravitational field $\boldsymbol{g}$ by the electric field $\boldsymbol{E}$ and the constant G by $-1/4\pi\varepsilon_0$.

(b) Capacity of a spherical capacitor

As an example of the result derived above we calculate the capacity of a spherical capacitor with internal and external radii r_1 and r_2. The capacitor is shown in figure 5.9.

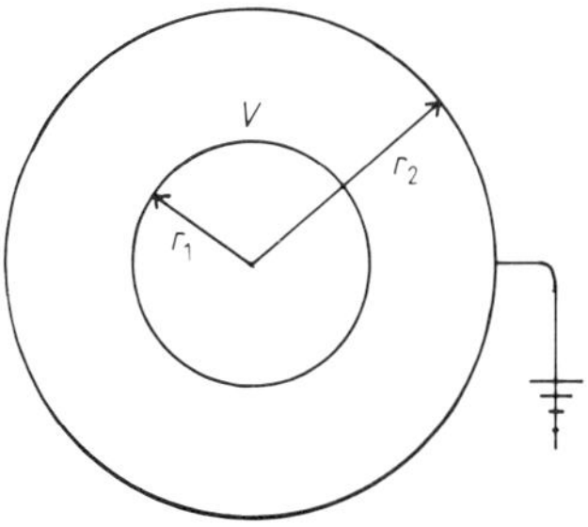

Figure 5.9

If the inside sphere is maintained at a voltage (potential) V and the outside sphere is earthed, there can be no field outside the capacitor (since there is no potential gradient there). Therefore, from Gauss's law the total charge on the two spheres must be zero. This means that there must be a charge Q on the inner spheres and a charge $-Q$ on the outer sphere. Between the two spheres we can use the result of example (*a*) to deduce that the electric field will be that due to a charge Q at the origin, i.e. $\boldsymbol{E}(r) = (Q/4\pi\varepsilon_0 r^2)\hat{\boldsymbol{r}}$, or $\varphi(r) = -Q/4\pi\varepsilon_0 r$. The voltage V is just the difference between the potential at r_1 and r_2:

$$V = \frac{Q}{4\pi\varepsilon_0}\left(\frac{1}{r_1} - \frac{1}{r_2}\right).$$

This gives us the capacity $C = Q/V$ as

$$C = \frac{4\pi\varepsilon_0 r_1 r_2}{r_2 - r_1}. \tag{5.21}$$

(c) Resistance of a cylindrical conductor

A resistor consists of a medium of resistivity ρ between two cylinders of radii r_1 and r_2 of length L as in figure 5.10.

If a voltage V is applied between the inner and outer cylinders a current I will flow. By symmetry the current density $\boldsymbol{j}$ will be radial, $\boldsymbol{j}(\boldsymbol{r}) = j(r)\hat{\boldsymbol{r}}$. The flux of $\boldsymbol{j}$ through any cylindrical surface of radius r, between r_1 and r_2, will be equal to the total current I:

$$\oint \boldsymbol{j}\cdot \mathrm{d}\boldsymbol{s} = I.$$

Because $\boldsymbol{j}$ is always parallel to the surface element $\mathrm{d}\boldsymbol{s}$ and of constant magnitude $j(r)$ the integral can be done immediately to give

$$I = 2\pi r L j(r). \tag{5.22}$$

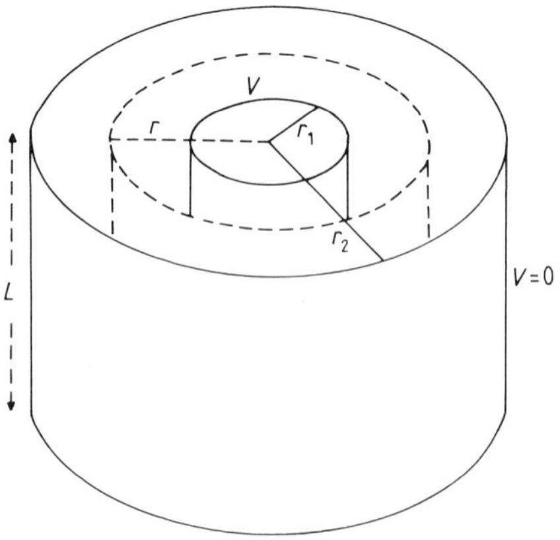

Figure 5.10

We now use Ohm's law in the form of equation (4.20)

$$\boldsymbol{j} = -\sigma\nabla\varphi = -\rho^{-1}\nabla\varphi$$

and carry out the line integral of $\boldsymbol{j}$ from radius r_1 to r_2:

$$\int \boldsymbol{j}\cdot \mathrm{d}\boldsymbol{r} = -\int \frac{1}{\rho}\nabla\varphi\cdot \mathrm{d}\boldsymbol{r} = -\frac{1}{\rho}\int \mathrm{d}\varphi$$

where we have used equation (4.13). Substituting the value of $j(r)$ from equation (5.22) and remembering that it is radial we get

$$\int_{r_1}^{r_2} \frac{1}{2\pi r L}\mathrm{d}r = -\int_{r_1}^{r_2} \frac{\mathrm{d}\varphi}{\rho} = -(\varphi(r_1) - \varphi(r_2))/\rho = V/\rho$$

i.e.

$$\frac{I}{2\pi L}\ln\left(\frac{r_2}{r_1}\right) = \frac{V}{\rho}$$

or

$$I = \frac{2\pi L V}{\rho \ln(r_2/r_1)}. \tag{5.23}$$

This means that the resistance is

$$R = \frac{V}{I} = \frac{\rho \ln(r_2/r_1)}{2\pi L}. \tag{5.24}$$

Because of the similarity of electrical conductivity, thermal conductivity and diffusion as seen in equations (4.20), (4.22) and (4.23), the same method could be used to derive the heat current or diffusion current in situations with the same geometry.

The Curl of a Vector Field 6

As in the last chapter we can formally define the curl by saying that curl $\boldsymbol{F}(\boldsymbol{r})$ is the vector field produced by taking the vector product of the vector operator ∇ with $\boldsymbol{F}(\boldsymbol{r})$ i.e.

$$\operatorname{curl}\boldsymbol{F}(\boldsymbol{r}) = \nabla\wedge\boldsymbol{F}(\boldsymbol{r}). \tag{6.1}$$

We could then write out Cartesian components of this equation but since this would shed little light on the meaning of the curl we go directly to the more intuitive form for the curl obtained from a line integral. We consider a small element of surface area $\mathrm{d}\boldsymbol{s} = \mathrm{d}s\hat{\boldsymbol{n}}$ and calculate the line integral $\boldsymbol{F}(\boldsymbol{r})\cdot\mathrm{d}\boldsymbol{r}$ around its perimeter. The value of this integral divided by the area $\mathrm{d}s$, in the limit as the area tends to zero, will turn out to be the component of curl $\boldsymbol{F}(\boldsymbol{r})$ in the direction $\hat{\boldsymbol{n}}$, i.e.

$$(\operatorname{curl}\boldsymbol{F})\cdot\hat{\boldsymbol{n}} = (\nabla\wedge\boldsymbol{F})\cdot\hat{\boldsymbol{n}}$$

$$= \lim_{\mathrm{d}s\to 0}(\text{line integral } \boldsymbol{F}(\boldsymbol{r})\cdot\mathrm{d}\boldsymbol{r} \text{ around } \mathrm{d}s\hat{\boldsymbol{n}})/\mathrm{d}s \tag{6.2}$$

$$= \lim_{\mathrm{d}s\to 0}\frac{1}{\mathrm{d}s}\oint\boldsymbol{F}(\boldsymbol{r})\cdot\mathrm{d}\boldsymbol{r}. \tag{6.3}$$

At the moment there is some ambiguity about the direction of $\hat{\boldsymbol{n}}$ and the direction of travel for the line integral. This is removed by requiring the line integral to be taken round the curve in the sense of a positive right-hand screw about the direction of $\hat{\boldsymbol{n}}$. We show that the result (6.3) is equivalent to the definition (6.1) by considering the small element of area to be a rectangle in the x–y plane with sides from x to $x + \mathrm{d}x$ and y to $y + \mathrm{d}y$ as in figure 6.1. The line integral around the perimeter is

$$\oint\boldsymbol{F}\cdot\mathrm{d}\boldsymbol{r} = \int_x^{x+\mathrm{d}x} F_x(x, y, z)\,\mathrm{d}x + \int_y^{y+\mathrm{d}y} F_y(x + \mathrm{d}x, y, z)\,\mathrm{d}y$$
$$+ \int_{x+\mathrm{d}x}^{x} F_x(x, y + \mathrm{d}y, z)\mathrm{d}x + \int_{y+\mathrm{d}y}^{y} F_y(x, y, z)\,\mathrm{d}y.$$

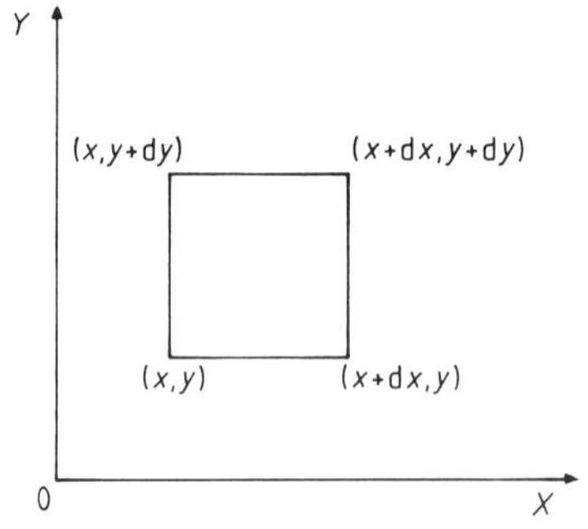

Figure 6.1

The first and third terms give a contribution

$$[F_x(x, y, z) - F_x(x, y + \mathrm{d}y, z)]\mathrm{d}x = -\left(\frac{\partial}{\partial y}F_x(x, y, z)\mathrm{d}y\right)\mathrm{d}x$$

since we can write $F_x(x, y + \mathrm{d}y, z) = F_x(x, y, z) + (\partial F_x(x, y, z)/\partial y)\mathrm{d}y$. In the same way the second and fourth terms give a contribution

$$\frac{\partial}{\partial x}F_y(x, y, z)\mathrm{d}x\mathrm{d}y.$$

The complete line integral around the rectangle is thus

$$\oint \boldsymbol{F}(\boldsymbol{r})\cdot\mathrm{d}\boldsymbol{r} = \left(-\frac{\partial F_x}{\partial y} + \frac{\partial F_y}{\partial x}\right)\mathrm{d}x\mathrm{d}y$$

and

$$\lim_{\mathrm{d}s\to 0}\frac{1}{\mathrm{d}s}\oint \boldsymbol{F}(\boldsymbol{r})\cdot\mathrm{d}\boldsymbol{r} = \frac{1}{\mathrm{d}x\mathrm{d}y}\left(-\frac{\partial F_x}{\partial y} + \frac{\partial F_y}{\partial x}\right)\mathrm{d}x\mathrm{d}y$$

$$= \frac{\partial F_y}{\partial x} - \frac{\partial F_x}{\partial y} = (\nabla\wedge\boldsymbol{F})\cdot\hat{\boldsymbol{k}} \qquad (6.4)$$

which is the result (6.3) for $\hat{\boldsymbol{n}}$ in the z-direction. We have made the same sort of approximations in deriving this result that we made in §5.4 and again they can be justified and the result shown to be independent of the shape and orientation of the small surface element.

6.1 Stokes's Theorem

This theorem for curl $\boldsymbol{F}$ is the analogue of Gauss's theorem for div $\boldsymbol{F}$ derived in §5.5. The theorem is

$$\int_S \mathrm{curl}\,\boldsymbol{F}\cdot\mathrm{d}\boldsymbol{s} = \oint \boldsymbol{F}\cdot\mathrm{d}\boldsymbol{r} \qquad (6.5)$$

where the line integral on the right is over the boundary of the surface on the left. It follows directly from the form for curl $\boldsymbol{F}$ used in equation (6.3) if we divide the surface into a network of small surface elements $\mathrm{d}\boldsymbol{s}_i$ as in figure 6.2. For each surface element equation (6.3) gives

$$(\mathrm{curl}\,\boldsymbol{F})\cdot\hat{\boldsymbol{n}}\,\mathrm{d}s_i = (\mathrm{curl}\,\boldsymbol{F})\cdot\mathrm{d}\boldsymbol{s}_i = \int_{\mathrm{d}s_i} \boldsymbol{F}\cdot\mathrm{d}\boldsymbol{r}$$

in the limit as $\mathrm{d}\boldsymbol{s}_i$ goes to zero. Summing over all elements $\mathrm{d}\boldsymbol{s}_i$ and taking the limit, the left-hand side becomes $\int \mathrm{curl}\,\boldsymbol{F}\cdot\mathrm{d}\boldsymbol{s}$. On the right-hand side the line integrals coming from two adjacent cells over the common line dividing them cancel since we traverse the line in opposite directions (see figure 6.2). Thus the only line elements which give a non-vanishing contribution are those over the boundaries of the surface $\boldsymbol{S}$. These, when summed, form the line integral round the boundary of the surface $\boldsymbol{S}$, i.e. the right-hand side of equation (6.5) and the result is proved. We note that in using equation (6.5) the direction of the surface normal and the sense in which the line integral is taken must be related as described immediately after equation (6.3). Of course the direction of $\hat{\boldsymbol{n}}$ varies as we go over the surface but its sense can be fixed by considering any small element of $\boldsymbol{S}$ touching the boundary curve.

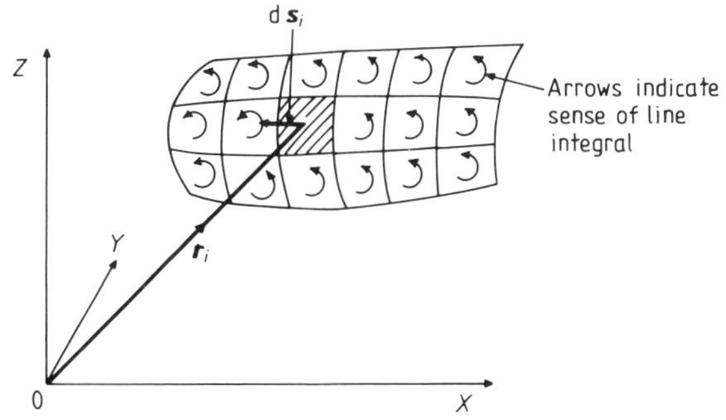

Figure 6.2

6.2 Interpretation of curl $\boldsymbol{F}$

In order to get a feeling for the meaning of curl $\boldsymbol{F}$ we will look at a field $\boldsymbol{F}$ which will turn out to have a constant value for curl $\boldsymbol{F}$. Such a field is provided by the velocity field for a rigid body in uniform rotation about a fixed axis. We can do this in a purely formal way starting from the definition (6.1).

The velocity of the point $\boldsymbol{r}$ in a body rotating with angular velocity vector $\boldsymbol{\omega}$

about an axis through the origin was given in equation (3.22)

$$\boldsymbol{v}(\boldsymbol{r}) = \boldsymbol{\omega}\wedge\boldsymbol{r}$$

and this is the velocity field we consider. First, using (6.1) we have

$$\text{curl}\,\boldsymbol{v} = \nabla\wedge\boldsymbol{v} = \nabla\wedge(\boldsymbol{\omega}\wedge\boldsymbol{r}).$$

We now work in Cartesian components and look just at one component

$$\begin{aligned}(\text{curl}\,\boldsymbol{v})_x = (\nabla\wedge(\boldsymbol{\omega}\wedge\boldsymbol{r}))_x &= \frac{\partial}{\partial y}(\boldsymbol{\omega}\wedge\boldsymbol{r})_z - \frac{\partial}{\partial z}(\boldsymbol{\omega}\wedge\boldsymbol{r})_y \\ &= \frac{\partial}{\partial y}(\omega_x y - \omega_y x) - \frac{\partial}{\partial z}(\omega_z x - \omega_x z) \\ &= \omega_x + \omega_x = 2\omega_x.\end{aligned}$$

A similar result holds for the other components so we have

$$\text{curl}\,\boldsymbol{v} = 2\boldsymbol{\omega}. \tag{6.6}$$

We have thus shown that for uniform rotational motion about an axis, curl $\boldsymbol{v}$ is just twice the angular velocity vector and is constant throughout. A field $\boldsymbol{F}$ for which curl $\boldsymbol{F}$ vanishes everywhere is therefore said to be 'irrotational'. The value of curl $\boldsymbol{F}$ is sometimes called the 'vorticity' a word with more obvious meaning if we take $\boldsymbol{F}$ to be the velocity field in a fluid.

We illustrate some fields which have vanishing and non-vanishing curl (or div) in the following figures.

(*a*) A constant field in the x-direction

$$\boldsymbol{F} = a\,\hat{\boldsymbol{i}}$$
$$\text{div}\,\boldsymbol{F} = 0$$
$$\text{curl}\,\boldsymbol{F} = 0$$

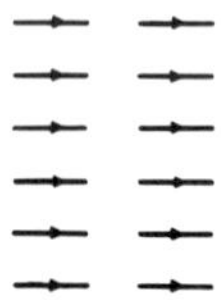

(*b*) A radial field

$$\boldsymbol{F} = a\boldsymbol{r}$$
$$\text{div}\,\boldsymbol{F} = 3a$$
$$\text{curl}\,\boldsymbol{F} = 0$$

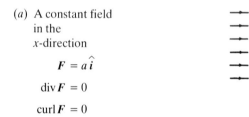

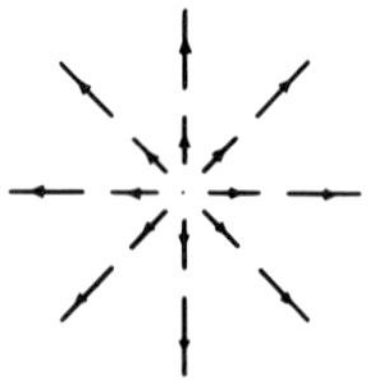

(*c*) A uniformly rotating field

$$\boldsymbol{F} = \boldsymbol{a} \wedge \boldsymbol{r}$$

$$\operatorname{div} \boldsymbol{F} = 0$$

$$\operatorname{curl} \boldsymbol{F} = 2a\hat{\boldsymbol{k}}$$

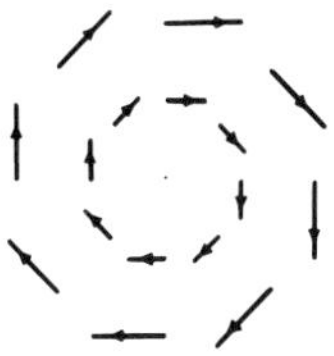

(*d*) A 'spiral' field

$$\boldsymbol{F} = \boldsymbol{a} \wedge \boldsymbol{r} + b\boldsymbol{r}$$

$$\operatorname{div} \boldsymbol{F} = 3b$$

$$\operatorname{curl} \boldsymbol{F} = 2a\hat{\boldsymbol{k}}$$

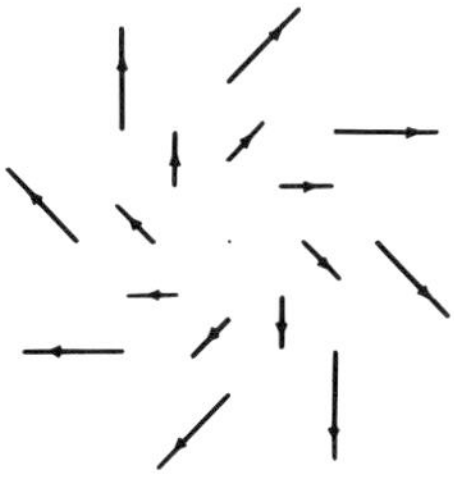

(*e*) A field growing in the *x*-direction

$$\boldsymbol{F} = ax\hat{\boldsymbol{i}}$$

$$\operatorname{div} \boldsymbol{F} = a$$

$$\operatorname{curl} \boldsymbol{F} = 0$$

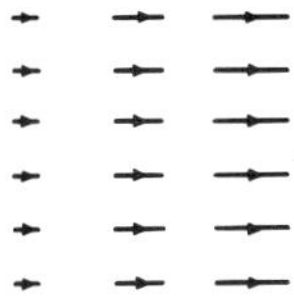

(*f*) A field growing in the *y*-direction

$$\boldsymbol{F} = ay\hat{\boldsymbol{i}}$$

$$\operatorname{div} \boldsymbol{F} = 0$$

$$\operatorname{curl} \boldsymbol{F} = -a\hat{\boldsymbol{k}}$$

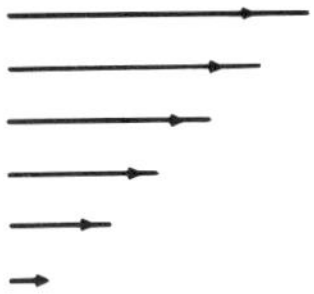

In these figures the magnitude and direction of the field at a given point are represented by a vector drawn with its 'tail' at that point. (This is not the usual representation of fields by flux lines with the density of lines representing the field strength.)

It is perhaps helpful when picturing the curl of a field $\boldsymbol{F}$ to imagine a small tennis ball immersed in the field and to think of the field as representing a force field on the ball. If the effect of the field is to make the ball rotate then the field has a non-vanishing curl and the direction of curl $\boldsymbol{F}$ is along the axis of rotation.

6.3 Double Vector Operators

In this section we consider the effect of operating twice with the ∇ operator on scalar and vector fields.

(a) curl(grad V) = $\nabla\wedge(\nabla V)$

It is easily shown, by writing out the Cartesian components that

$$\text{curl}(\text{grad}\, V) = 0. \tag{6.7}$$

Since the electrostatic and gravitational fields can be written as gradients of scalar potential fields (as in §4.3) it follows that

$$\text{curl}\,\boldsymbol{E} = 0 \qquad \text{curl}\,\boldsymbol{g} = 0.$$

These fields are therefore irrotational. In fact it can be shown that the necessary condition for a vector field to be the gradient of a scalar field is that it be irrotational.

(b) div(curl F) = $\nabla\cdot(\nabla\wedge\boldsymbol{F})$

Again this can be shown to vanish identically so

$$\text{div}(\text{curl}\,\boldsymbol{F}) = 0. \tag{6.8}$$

In §5.5 we pointed out that $\text{div}\,\boldsymbol{B} = 0$ for the magnetic field $\boldsymbol{B}$. Such a field is said to be 'solenoidal'. From equation (6.8) we see that we can write a solenoidal field as the curl of another vector field, usually called the vector potential field. In the case of the magnetic field $\boldsymbol{B}$ this is the magnetic vector potential $\boldsymbol{A}$, i.e.

$$\boldsymbol{B} = \text{curl}\,\boldsymbol{A}. \tag{6.9}$$

This equation does not determine $\boldsymbol{A}$ uniquely, but it can be shown that $\boldsymbol{A}$ is determined if $\text{div}\,\boldsymbol{A}$ is also specified. It is usual to choose $\text{div}\,\boldsymbol{A} = 0$.

(c) div(grad V) = $\nabla\cdot(\nabla V) = \nabla^2 V$

If this operator is written out in Cartesian components it is found to be

$$\text{div}(\text{grad}\, V) = \frac{\partial^2 V}{\partial x^2} + \frac{\partial^2 V}{\partial y^2} + \frac{\partial^2 V}{\partial z^2} \tag{6.10}$$

and this is written as $\nabla^2 V$ where the operator

$$\nabla^2 = \frac{\partial^2}{\partial x^2} + \frac{\partial^2}{\partial y^2} + \frac{\partial^2}{\partial z^2}$$

is called the Laplacian operator.

For electric fields we can write

$$\nabla^2\varphi = \text{div}(\text{grad}\,\varphi) = -\text{div}\,\boldsymbol{E} = -\rho/\varepsilon_0$$

which is Poisson's equation. If $\rho = 0$ then this becomes Laplace's equation

$$\nabla^2\varphi = 0.$$

(d) $\nabla^2\boldsymbol{F}$

The Laplacian operator can also act on a vector field $\boldsymbol{F}$ with the result

$$\nabla^2 \boldsymbol{F} = \frac{\partial^2 \boldsymbol{F}}{\partial x^2} + \frac{\partial^2 \boldsymbol{F}}{\partial y^2} + \frac{\partial^2 \boldsymbol{F}}{\partial z^2} \tag{6.11}$$

which can also be written in the alternative form

$$\nabla^2 \boldsymbol{F} = \hat{\boldsymbol{i}}\,\nabla^2 F_x + \hat{\boldsymbol{j}}\,\nabla^2 F_y + \hat{\boldsymbol{k}}\,\nabla^2 F_z$$

(e) curl(curl $\boldsymbol{F}$*)* $= \nabla \wedge (\nabla \wedge \boldsymbol{F})$

By expanding in component form this can be shown to give

$$\operatorname{curl}(\operatorname{curl}\boldsymbol{F}) = \operatorname{grad}(\operatorname{div}\boldsymbol{F}) - \nabla^2 \boldsymbol{F}. \tag{6.12}$$

6.4 Examples Involving Curl

Some of the best examples of the use of the results of this chapter are in electromagnetic theory and we illustrate this by deriving one of the remaining two Maxwell equations (the first two are equations (5.14) and (5.17). We refer the reader to the book 'Maxwell's Equations and their Applications' by Thomas and Meadows in this same series for many further examples).

Faraday's law for the potential difference induced in a circuit by a varying magnetic field is $V = -\mathrm{d}\Phi/\mathrm{d}t$, where Φ is the magnetic flux though the circuit, i.e. $\Phi = \int \boldsymbol{B}\cdot \mathrm{d}\boldsymbol{s}$.

The potential difference induced along a small section $\mathrm{d}\boldsymbol{r}$ of the circuit will be given in terms of the electric field $\boldsymbol{E}$ as $\boldsymbol{E}\cdot\mathrm{d}\boldsymbol{r}$. We can therefore write

$$V = \oint_{\mathrm{C}} \boldsymbol{E}\cdot \mathrm{d}\boldsymbol{r} = -\frac{\partial}{\partial t}\int_{\mathrm{C}} \boldsymbol{B}\cdot \mathrm{d}\boldsymbol{s}$$

where the line integral is around the circuit C and the surface integral is over any surface bounded by C. Using the Stokes's theorem (6.5) this becomes

$$\int_{\mathrm{C}} \operatorname{curl}\boldsymbol{E}\cdot \mathrm{d}\boldsymbol{s} = -\int_{\mathrm{C}} \frac{\partial \boldsymbol{B}}{\partial t}\cdot \mathrm{d}\boldsymbol{s}$$

or

$$\int_{\mathrm{C}} \left(\operatorname{curl}\boldsymbol{E} + \frac{\partial \boldsymbol{B}}{\partial t}\right)\cdot \mathrm{d}\boldsymbol{s} = 0.$$

Since this holds for any circuit C the integrand itself must vanish. i.e.

$$\operatorname{curl}\boldsymbol{E} = -\frac{\partial \boldsymbol{B}}{\partial t}.$$

This is the third Maxwell equation.

The corresponding equation for magnetic fields is

$$\operatorname{curl}\boldsymbol{B} = \mu_0\left(\boldsymbol{j} + \varepsilon_0 \frac{\partial \boldsymbol{E}}{\partial t}\right).$$

This can be derived starting from the experimental Biot–Savart law (see for instance Thomas and Meadows, loc. sit.).

Index